William Lucas Distant

A Naturalist in the Transvaal

William Lucas Distant

A Naturalist in the Transvaal

ISBN/EAN: 9783337025502

Printed in Europe, USA, Canada, Australia, Japan

Cover: Foto ©berggeist007 / pixelio.de

More available books at **www.hansebooks.com**

[Frontispiece.

Makapan's Cave.

A NATURALIST IN THE TRANSVAAL.

BY

W. L. DISTANT,

MEMBER OF THE ANTHROPOLOGICAL INSTITUTE, THE ENTOMOLOGICAL SOCIETY OF FRANCE,
FELLOW OF THE ENTOMOLOGICAL SOCIETY OF LONDON,
AND CORRESPONDING MEMBER OF THE ENTOMOLOGICAL SOCIETY OF STOCKHOLM
AND OF THE BUFFALO SOCIETY OF NATURAL SCIENCES.

WITH COLOURED PLATES AND ORIGINAL ILLUSTRATIONS.

"Nature retains her veil, despite our clamours:
That which she doth not willingly display
Cannot be wrenched from her with levers, screws, and hammers."
—'Faust' (*Bayard Taylor's Transl.*).

LONDON:
R. H. PORTER, 18 PRINCES STREET, CAVENDISH SQUARE, W.
1892.

PRINTED BY TAYLOR AND FRANCIS,
RED LION COURT, FLEET STREET.

TO THE MEMORY OF MY FATHER,

Alexander Distant,

WHO, IN OLD SOUTH-SEA WHALING-DAYS, SAILED
ROUND AND ROUND THE WORLD,
AND TRANSMITTED A LOVE OF ROAMING TO HIS SONS,

I DEDICATE THIS BOOK.

PREFACE.

The following pages record the impressions of a naturalist, who, during a twelve months' business sojourn in the Transvaal, deprived of the society of family and friends, employed the whole of his leisure time in that most delightful consolation—zoological recreation.

In my schoolboy days a journey through the Transvaal would have almost attained the dignity of an exploration; now Pretoria can be reached in three weeks' time from London, and the once long wagon-trek from the Cape is replaced by less than two days' train and a little more than two days' coach service. But this facility of transit, so valued by the business man and so necessary to the material development of the country, has deprived the sportsman of a hunting-ground and curtailed the view of the naturalist. No longer do vast herds of ruminants roam over these solitary plains, for when commerce reached the land, and bid for the skins of the buck and antelope, the Boer accepted the price and slaughtered, if not actually exterminated, the finest ruminant fauna that ever a land possessed. Further inland the Kafir, armed with a gun, pursues the same desultory warfare, and this portion of Southern Africa has completely lost what was once its most distinctive zoological feature. This animal extinction has also reacted on the Boer himself: now no longer the mighty hunter, he will soon cease to be the matchless marksman as of old; and his life on the solitary farm is thereby rendered more monotonous, for the gun was once his constant companion. When railways intersect the country the ox-wagon will gradually disappear, and with it the last characteristic feature of the old "voortrekkers."

The Transvaal is thus changed in its natural aspects from a tract once supporting an immense number of wild animals, and peopled by rugged farmers who lived a semi-pastoral, semi-hunting existence, to a country becoming progressively subject to European laws and customs, in which the earlier rough struggle for existence is now transformed into a race for wealth. The lawyer and the financier thrive where in recent years the lion and leopard fought for food, and townships have sprung up on spots where living Boers have formerly shot big game.

I thus saw the old order changing, and a state basing its progress solely on the foundation of auriferous reefs, for the future of the Transvaal largely depends upon the development of its mineral wealth. But the real Boer population form no appreciable portion of the inhabitants which reside in the large towns and depend on commerce and mining; the true Boer is still a farmer, and a few high officials do not adequately represent the characteristics of what—let alone—would have formed a distinct race of Dutch people. I hope I have been fair with these emigrant farmers, whom I really respect, but it is difficult to steer clear of both Boer and British prejudice: the first resents any criticism, the second criticises in a too sweeping and trenchant manner.

In an Appendix I have given an enumeration of my zoological collections, which were much assisted by an old and valued servant, Timothy Donovan, who accompanied me to the Transvaal. The proportion of new species is perhaps as high as might have been expected from the number of specimens collected, which may provide the material by which to commence a tabulation of the fauna of the Pretoria District, and also show that even a busy man, during his leisure hours, may do some not altogether useless biological work.

The lamented death of my friend, Mr. H. W. Bates, as these pages were passing through the press, adds a melancholy reflection to the obligations I am under, for his reading of my proofs with valuable suggestions. These were probably the last of the many friendly offices he undertook to aid his favourite study and to oblige his friends.

To the specialist friends who have aided me in working out my collections I render my best thanks, and have individually acknowledged their kind help when enumerating the different Orders in the Appendix. My travelling companion, Mr. Henry Blackwell, Jun., who allowed the use of some photographs taken by himself, has placed me under great obligation, and I am also indebted to the painstaking care of my artists, Mr. P. J. Smit and Mr. W. Purkiss. Last, but not least, my thanks are due to my publisher for having afforded me every facility to produce this small book in a worthy guise and manner.

Purley, Surrey.
February, 1892.

CONTENTS.

	PAGE
DEDICATION	v
PREFACE	vii
LIST OF ILLUSTRATIONS	xv

CHAPTER I.

TO PRETORIA.

Sail for South Africa.—Passengers illustrate evolutionary factors in the formation of a Colony.—Zoological observations at sea.—Flying-fish.—Malays at Cape Town.—South-African Museum.—Port Elizabeth.—Different routes to the Transvaal.—Durban.—Railway views between Durban and Newcastle.—Coach-travelling and its incidents.—Majuba Hill and scenes of late Boer War.—Extermination of the ruminant-fauna.—Johannesburg after the boom.—Pretoria; botanical features; design of the town 1

CHAPTER II.

THE BOER.

Where are the Boers?—The Boer a farmer.—Grass-fires and their consequences.—Habits of the farmer.—Peculiar theology of the Boer which governs his life and action.—Boer relations to the Kafirs.—Violence of Church disputes.—President Krüger.—Some causes of the Boer War.—The Boers as soldiers.—Homely life of the President; his great influence with the Boers.—Many farmers now wealthy men.—Physical characteristics of the Boers; their supposed dislike to the British; their mistrust of the Hollanders 20

CHAPTER III.

PHASES OF NATURE AROUND PRETORIA.

PAGE

Natural aspects in the dry winter season.—Orthoptera and Coleoptera.—Commencement of the rainy season.—Protective resemblance in butterflies.—Vegetable tanning-products.—Survival of spined and hard-wooded trees in the struggle for existence with herbivorous fauna.—Baboons.—Bad roads.—A Boer farm.—Grass-fires.—Dust-storm.—Vast quantities of beetles under stones.—Bad weather and heavy losses in live stock.—Appearance of winged Termites.—Swollen streams and their dangers.—Accidental dangers in animal life.—Birds of Prey 38

CHAPTER IV.

PHASES OF NATURE AROUND PRETORIA (continued).

Geological features.—Dendritic markings.—The highlands and the sea.—Heavy rains and floods.—A protected butterfly and its enemy.—Mimicry.—Cicadas.—Species found both in England and the Transvaal.—The Secretary-bird.—Vultures.—Locust-swarm.—The Paauw and other Bustards.—The Monitor.—Partridges.—Evolution and struggle for existence 58

CHAPTER V.

THROUGH WATERBERG.

Scarcity of timber in the Transvaal.—Leave Pretoria for Waterberg.—Waterless region of the Flats.—The Warm Baths.—Beautiful scenery.—Euphorbias and their poisonous qualities.—Fever districts.—The Massacre at Makapan's Poort.—Sanguinary retribution at Makapan's Cave.—A fine orthopterous insect.—The Prospector.—Reptiles.—Ravages of the "Australian Bug."—Majuba day.—Mimicking insects 77

CHAPTER VI.

ZOUTPANSBERG AND THE MAGWAMBAS.

Start for the Spelonken in Zoutpansberg.—Horse-sickness.—Pietersburg.—A fine Convolvulus.—A castellated residence in the Wilds.—Night in a wagon.—Kafir traders.—Kafirs on the tramp.—Polygamy.—The Magwambas, their customs and institutions.—An ox feast and dance.—The Makatese.—The Mavendas and their ironwork.—Birds' food largely orthopterous.—Good entomological spots.—Zoutpansberg with its natural riches still undeveloped 94

CHAPTER VII.

A JOURNEY TO DURBAN.

Acacia mollissima.—Heavy cost on imports to the Transvaal.—Johannesburg and its Hotels.—Heidelberg.—A Priest of Islam.—Across the Ingogo heights to Newcastle.—Durban.—Colonel Bowker.—Best collecting-grounds around Durban.—Flowers, fruit, and insects.—Peculiarities in railway construction.—Model Natal farms.—Insect-pests to gardens.—Difficulties in coaching after heavy rains.—The store- and canteen-keeper of the veld 115

CHAPTER VIII.

THE MEN OF PRETORIA.

The inhabitants of Pretoria.—Auriferous wealth alone the present cause of Transvaal development.—Uneducated condition of the Boers.—Liquor traffic with the Kafirs.—The British colonist in the Transvaal.—The Hebrew in Pretoria.—Commercial morality.—The name of Mr. Gladstone execrated in the Transvaal.—The Kafir and his value as a labourer.—Sanitary condition of Pretoria.—Vital statistics.—After-effects of the boom.—Attachment of Colonists to their adopted country 132

APPENDIX.

                                                                             PAGE

Enumeration and Description of the Anthropological and Zoological Objects collected by the Author, with Contributions by ERNEST E. AUSTEN, Zool. Dept. Brit. Mus., G. A. BOULENGER, Zool. Dept. Brit. Mus., F.Z.S., JULES BOURGEOIS, M.E.Soc.Fr., J. H. DURRANT, F.E.S., C. J. GAHAN, M.A., Zool. Dept. Brit. Mus., Rev. H. S. GORHAM, F.Z.S., MARTIN JACOBY, F.E.S., R. I. POCOCK, Zool. Dept. Brit. Mus., H. DE SAUSSURE, Socius hon. Soc. ent. Lond. Rossic. Belg., &c., &c., Prof. C. STEWART, Pres. Linn. Soc., &c., &c., and the AUTHOR 151

MAMMALIA 157

AVES 163

REPTILIA AND BATRACHIA 173

ARACHNIDA AND MYRIOPODA 179

INSECTA 187

INDEX 263

LIST OF ILLUSTRATIONS.

	PAGE
Makapan's Cave	*Frontispiece*
Good-bye to the Tender	1
Pier-head, Cape Town	4
Changing Mules on the Veld	10
View in the Town of Pretoria	15
Boer Wagon with Firewood	20
President Krüger	29
Chera progne. Male in nuptial plumage	38
Batrachians devouring Termites	49
After the Rains. Coleoptera	51
Buteo desertorum. Post of Observation	56
Dendritic Markings in Quartzite	58
Hemisaga prædatoria, n. sp.	63
Locust-swarm in Pretoria	71
The Monitor (*Varanus niloticus*)	77
Clonia wahlbergi	83
Native Hut, Spelonken	94

	PAGE
Castellated Residence in Zoutpansberg	97
Magwamba Woman crushing Meal	101
Native Arts of the Spelonken . . . . *facing*	102
Magwamba War-Axes	103
Magwamba Assegais and Shield	105
Native Iron-smelting	109
Magwamba Carvings	114
Apple-destroyers in Natal	115
Mylabris transversalis on Rose	127
Kafir Shepherd	132
Native Policeman	141
Glauconia distanti	175
Spirostreptus transvaalicus	182

COLOURED PLATES:
    Tab. I. New Species of Coleoptera.
        II. ,, Rhynchota.
        III. Lepidoptera and Rhynchota.
        IV. Orthoptera, Lepidoptera, Hymenoptera, and Coleoptera.

UNCOLOURED PLATE:
    Tab. V. New Species of Arachnida and Hymenoptera.

GOOD-BYE TO THE TENDER.

CHAPTER I.

TO PRETORIA.

Sail for South Africa.—Passengers illustrate evolutionary factors in the formation of a Colony.—Zoological observations at sea.—Flying-fish.—Malays at Cape Town.—South-African Museum.—Port Elizabeth.—Different routes to the Transvaal.—Durban.—Railway views between Durban and Newcastle.—Coach-travelling and its incidents.—Majuba Hill and scenes of late Boer War.—Extermination of the ruminant-fauna.—Johannesburg after the boom.—Pretoria; botanical features; design of the town.

On a fine June day in 1890, the 'Norham Castle' slowly hauled through the Dock-gates and steamed down the river, to carry many hopes and fears to South Africa. At Dartmouth our principal contingent of passengers joined the vessel and we sped South. It is a well-established rule that readers shall be spared the dreary recital of a voyage that has now lost all its romance with increased speed and additional comfort, though a floating hotel was a strong contrast to the surroundings of my trip to the Malay Peninsula in a sailing-vessel twenty-three years previously.

The sociologist, however, may see much to interest and instruct him in the curious group of individualities which constitute the passengers on an ocean steamship. Thrown so close together, with no outside news of the world, we notice each other's peculiarities and expose our own. South Africa is now not only a health resort for the invalid, and a campaigning ground for the commercial traveller, but its gold-fields attract those spirits of enterprise and speculation who wait on fortune and scorn laborious days. By the side of the sufficiently opulent man of weak constitution who can afford the time and money incidental to a trip for health, is the commercial traveller who now carries his samples through the colonies as he once "worked" the United Kingdom, and starts for the Cape with little more preparation than he would have previously made for a journey to the North of England. Church of England curates—who take precedence on Sundays—and members of High Anglican Sisterhoods sail along with highly educated and less educated dissenting ministers. Forward our hardy mechanic, "whose bones were made in England," who will carry his handicraft, his energy, and also his love of "wholesome beer" to a colony that will be certainly the richer for his first two qualifications, rubs shoulders with the lower form of Israelite, who does not compliment his race, who may possibly buy "illicit diamonds," or even succeed to the greater height of assisting in the promotion of a bogus gold company. All, however, are "hail, good fellows, well met," on board, and though saloon, second-class, and steerage are a little timorous of each other afloat, the distinctions are not so accentuated as on shore. It is in these migratory assemblies that one may study the evolution of a colony.

There is little opportunity for a naturalist on board a fast steamer; and for one who has travelled the ocean before, the animals met are much the same. But after twenty years the sight of a flying-fish is a renewed delight. We first met with the genus—for there are several species—near Madeira, and here *Exocœtus** *lineatus*, the largest species, is found; further south the flying-fish are more numerous, but smaller in size. This fish is certainly the most prolific of any to be found in the warmer parts of the ocean, and its numbers are simply prodigious. All day, and presumably all night, as the vessel ploughs its course, it constantly disturbs and disperses the fish, which in these parts must universally populate the surface waters. I have frequently spent considerable time looking over the bows of vessels, and watched the almost incessant flight of the frightened creatures as the ship, like a monstrous enemy, tore through their midst. In recent years much attention has been paid to the question as to whether these *Exocætans* flap their wings during flight, or simply skim with expanded wings from the initial velocity with which they leave the water. My own observations certainly incline to the last opinion, and that the rising of the fish was coincident with a rapid movement of the tail, which always more or less reminded one of the action of the blade of a steamer's screw. This can also be frequently observed when the fish at the end of its flight apparently observes a lurking enemy†, and just touching the water, the action of its tail can be again noticed preparatory to a fresh escape from the sea. There can be little doubt that the flight of this fish is always of a protective character, and is scarcely undertaken for any pleasure or relaxation. A ship must appear as a hideous monster, and add to the many terrors in the lives of these abundant animals: some are more alarmed than others, for many of them will again take to the water but a short distance from the vessel,

---

* This generic name is derived from a curious belief of the ancients, who were under the impression that these "sea swallows," as they called the flying-fishes, left the ocean at night and slept on shore, in order to be free from their enemies. From this practice of "sleeping out" they were named *Exocœti*.

† Pigafetta witnessed this in 1520, and quaintly wrote: "Meanwhile their enemies follow their shadow, and arriving at the spot where they fall, seize upon them and devour them—a thing marvellous and agreeable to see."

whilst a few, and generally the larger specimens, pursue a more prolonged flight. Although I have never been able to detect any flapping of the wing-like pectoral fins of the flying-fish, I could not but agree with a previously recorded observation, that there appears to be a vibrating movement, as in the wings of a grasshopper*. The question, however, is dependent on the evidence of the senses and is difficult to determine. One very large school of Porpoises, an occasional Shark, basking or sleeping near the surface, many specimens of the "Portuguese Man-of-War" (*Physalia*) as we passed through the tropics, and oceanic birds (Mollymauks, Albatrosses, and Petrels) as we approached the cooler regions of the Cape, were all that gratified the greedy eye of a naturalist.

Pier-head, Cape Town.

We reached the Cape in fifteen days after leaving Madeira, at least a day late, and our impatience was

---

* For the argument and evidence against the non-flapping of the wings of the Flying-fish see Prof. Carl Möbius, "Die Bewegungen der fliegenden Fische durch die Luft" (Zeitschrift für wissenschaftliche Zoologie, Suppl. vol. xxx. p. 343, 1878). An opposite opinion has been advanced by C. O. Whitman ('American Naturalist,' vol. xiv. p. 641, 1880).

not calmed by the reflection that in the early days of discovery it took the Portuguese a hundred years with innumerable expeditions to double the same. Cape Town, with its thriving business community and its good shops, reminds one of a flourishing seaside town in England. The fishing-quarters are inhabited chiefly by Malays*, who seem, from long residence, to have quite lost the purity of their mother tongue, and the Malay women, in their best attire, affect a European costume, in which an enormous and hideous bloomer-skirt is the strongest point, a strange and unpleasant contrast to the graceful sarong I remembered in the Malay Peninsula. The South-African Museum, presided over by my old friend Roland Trimen, leaves nothing to be desired but greater space and more available funds for the acquisition of fresh specimens. One can form no adequate conception of the South-African fauna from the present compulsory crowded contents of this building. The arrangement of a museum should be the reflection of a man's grasp of Zoology, but a curator has no opportunity of displaying the same if sufficient space is not at his disposal. A local museum should perhaps follow the ideal of a man's knowledge, to know a little about everything, and everything about something; so it might be somewhat weak in several groups, but very strong and exhaustive in one particular branch of Natural History. This is the case here, for Mr. Trimen is a renowned lepidopterist, and the collection of butterflies is perhaps more complete and better worked out than can be found in any other of our colonial museums. One of its greatest treasures is the head of a "White Rhinoceros" (*Rhinoceros simus*). This now practically extinct mammal, which has been shot by living sportsmen, is unrepresented in any zoological menagerie, and its perfect skin or skeleton is unknown in any museum, thus affording a good illustration in the present day of

---

* The large body of Malay Mussulmans at the Cape have of late years come under the patronage of the Sultan. A school has also been founded at Kimberley by the Sultan, which, after him, has been named Hamidieh ('Athenæum,' Oct. 17, 1891).

how a species may disappear*. In South Africa more than one species of Buck and Antelope is rapidly approaching the same fate; and if it would be exaggeration to say the days, we may safely affirm that the years of the African Lion are numbered.

There are now five routes for reaching the Transvaal from South Africa. The first is from Cape Town direct:

	miles.
Cape Town to Kimberley (rail)	647
†Kimberley to Fourteen Streams (coach)	47
Fourteen Streams to Klerksdorp ,,	110
Klerksdorp to Potchefstroom ,,	31
Potchefstroom to Johannesburg ,,	80
Total miles	915

This is the quickest and favourite passenger route from England, and with it we may describe what is usually a heavy-goods route, and by way of Port Elizabeth:

	miles.
Port Elizabeth to Kimberley (rail)	485
Coach journey as detailed in Cape route	268
Total miles	753

The next route *viâ* Bloemfontein from Port Elizabeth, recently opened, is now being pushed to the Vaal River to meet the connection from Johannesburg; but before this can be done nearly a dozen either large rivers or spruits‡ have to be bridged over:

	miles.
Port Elizabeth to Colesberg (rail)	305
Colesberg to Norval's Pont (Orange River)	23
Norval's Pont to Bloemfontein (rail)	121
Bloemfontein to Johannesburg (coach)	250
Total miles	699

* In 'Nature,' vol. xlii. p. 520, Dr. Sclater has written on this matter and figured both the heads of *R. simus* and the common species *R. bicornis*.
† The rail now extends to Vryburg, by which the amount of coach-travelling is diminished.
‡ A "spruit" is a small stream or rivulet.

The East London line *via* Aliwal North to Johannesburg is the least used:

	miles.
East London to Aliwal North (rail)	280
Aliwal North to Johannesburg (coach)	320
Total miles	600

The Natal, and the most pleasant, route is by way of Durban:

	miles.
Durban (Point) to Charlestown (rail)	304
Charlestown to Johannesburg (coach) say	130
Total miles	434

The last way the Transvaal may be approached is *via* Delagoa Bay:

	miles.
Delagoa Bay to Moveni (rail)	62
Moveni to Barberton (coach)	98
Barberton to Pretoria (coach)	167
Pretoria to Johannesburg (coach)	32
Total miles	359*

I had business to transact at both Cape Town and Durban, so necessarily approached the Transvaal through Natal. The voyage between Cape Town and Durban, calling at Mossel Bay, Port Elizabeth, and East London, occupied a week. We reached Mossel Bay on Sunday morning, and the church service, which on previous Sundays had seemed to have been of very strict observance, was now suspended for the important operation of discharging cargo. Port Elizabeth has no claim to beauty, but possesses an exceedingly healthy climate, is renowned for having the most genial and hospitable community in South Africa, can justly be proud of its

---

* The accuracy of these figures can be relied upon, and they are extracted from some statistics specially prepared for the 'Johannesburg Standard and Digger's News,' by D. C. Stevens.

Botanic Gardens, which by Scotch industry and skill were made on the site of a sandy waste, and exhibits the most unsatisfactory local museum it was ever my lot to enter. A new curator is now engaged, and will probably remedy many of its present defects; but a commencement might be made by eliminating common Indian Lepidoptera, which are unnamed and unlocalized as such, and also by removing some of the brilliant paint of various hues by which it has been sought to ornament a Shark which hangs pendent from the roof. At East London rough weather prevented a comfortable landing, but here an ichthyologist would find much to interest him. Two Hammer-headed Sharks (*Zygana* sp.) patrolled the ship, whilst some of the crew threw out lines and caught Sea-Perch, Cape Salmon, and Dogfish. Both here and at Port Elizabeth sea-bathing is rendered dangerous by the presence of many large Sharks.

Durban, washed by the Indian Ocean, has a more or less Oriental aspect: gaily-dressed Klings walk the roads and show their old partiality for selling fruit and vegetables; it is the Hindu race that provide the railway porters and the hotel waiters, and a large number of the stores are kept by what are styled "coolie" merchants. Although it was still winter there was a warmth and colour about Durban that made the contrast to the Cape very pleasant and very tropical; but as Natal forms the subject of another chapter little need be said here, and our stay was very short. We landed at noon and left by the evening train for the Transvaal *viâ* Newcastle.

The railway passes through some of the finest scenery of Natal; but this part of the journey was completed during the night, and when daylight broke we were near Ladysmith, and mountain, ravine, and rivers were giving place to those bare and generally treeless tracts that are so universally known in South Africa as *veld*. From Ladysmith to Newcastle the rail ascends some steep inclines, which eventually lead to the high plateau on which Johannesburg and Pretoria stand, thus accounting for the temperate climate of that inner south-eastern portion of the continent. Scarcely a living thing could be seen from the carriage windows, the parched aspect of an African winter, which made the wilderness look more forlorn, was qualified by the clear light, the cloudless sky, and the pure dry but invigorating atmosphere. This railway is the main artery by which Natal carries on its large and increasing trade with the Transvaal. It is but a few years ago that Pietermaritzburg was the terminus, and from thence wagon and coach were the only further means of transport; then the iron way reached Ladysmith, afterwards pushed on to Biggarsberg, and at the time of our journey extended to Newcastle*, which we reached about midday. Biggarsberg particularly exhibits the migratory nature of these small termini. At the time when it represented—though but for a short period—the railway boundary, a very fair hotel was erected, large sheds were necessary to hold the merchandize that continually arrived and waited for wagon transport, whilst the neighbourhood became the residence of the different transport agents. Possessing nothing in itself, when the line extended to Newcastle, hotel, sheds, and transport agents passed on, and Biggarsberg to-day is a small village with a rather large railway station. Newcastle is in a different position, and although the carrying trade is now transferred to Charlestown, it possesses coal, and has a wool trade which will maintain its already somewhat advanced development.

It is singular to renew the old coaching days of England in South Africa, yet it is probable that the nearest approach to that method of travelling is to be found to-day in and near the Transvaal. We left Newcastle on a clear July Sunday noon, with a full load of twelve passengers, extra luggage (for each passenger is only allowed 28 lb.†), and the Natal mails, in a kind of open break with a team of eight horses. Of regular

---

* This was in 1890; in the spring of the year 1891 the line was opened for traffic as far as Charlestown, and now reaches the confines of the Transvaal.

† On my return journey by coach from Pretoria to Vryburg I was charged £7 extra for my trunk, although my personal passage was only £9 10s.

roads there are none, what we drove over are better described as good wide paths, like footways—but broader—across rugged common lands at home, with dips and hollows, large half-buried stones in some places, and small streams and rivulets—spruits—to cross occasionally, with jolting and bumping, which is the more noticeable on a first journey. But these rolling grassy plains and bare hills, stretching for hundreds of miles around, are not only invigorating, but positively exhilarating. It is winter, though the days are hot. No rain now falls, and the veld is covered with a close dried-up growth of herbage, giving a light brown tinge to the landscape till it meets the clear blue sky-line. It is at sunrise when these hues become intensified and tinged with the reflected solar light, and pale carmine and deep umber tints are then exhibited. We change horses about every hour at small wayside posting-houses, generally covered with the universal roof of corrugated iron, for here there are neither tiles nor slates, and wood has to be imported or transported to these treeless wastes.

CHANGING MULES ON THE VELD.

One man drives—" Cape Boys " excel at this work,—the conductor sits by his side, and it is he who wields the long whip and helps to pilot the driver. The road is up-hill, amidst mountains and glorious views ; Natal here bids her farewell to the Dutch Republic, and a wilderness again reigns beyond. We pass through the scene of the late Boer War, past Majuba Hill, and through Laings Nek : but it is a sorry subject ; all these fights took place on Natal territory which the

Boers had invaded, and brave English soldiers sleep around slain by the unerring bullets of plain Boer farmers who were held too cheap. Both sides were composed of brave men, but the rules of war observed by our commanders were too little marked by the subtlety of border warfare and too much by parade and field-day observance. Two small trees, since planted by his wife, mark the resting-place of the bold, genial, but unfortunate General Colley. These trees stand alone, the silence of the veld surrounds them; by Colley's side lies the body of a companion in arms, whilst Majuba Hill at a short distance frowns above. It is a bitter and a sad spot for Englishmen, and we feel relieved as the night covers us while passing through Laings Nek, and painful memories are left behind. Volksrust and a small posting-house or hotel is reached about 8 P.M., and now we have entered the Transvaal and our luggage is searched. The search is thorough, but courteous. Individuals who have lately had their word accepted by the Inspectors that they carried nothing excisable have afterwards boasted at Johannesburg and Pretoria how they have "done" the Customs and smuggled through their duty-paying effects; hence greater care is now taken and Englishmen have certainly no reason to complain. We take dinner and go to bed —always two and sometimes four beds in a room; but at 2.30 A.M. we are again aroused, and by 3 A.M. we are huddled up in the coach, for now the break is exchanged for the real mail-coach with a team of ten horses. It is perfectly dark and very cold, the windows are all pulled up, and though we have three ladies—who do not object—nine pipes are put in active work. One passenger tried very hard to start a conversation, but the darkness and the early hour were too depressing, and silence and tobacco resumed their sway. The dawn broke about 6 A.M., and a white frost was seen on the veld; but as the sun rose and the mists were dispelled the view once more asserted its lonely grandeur, the clear atmosphere became positively tonic, whilst a small herd of Buck were seen about a mile away. These animals are now rarely met, and then only in small numbers. Within quite recent years great herds were passed as one travelled through the country, and these plains actually swarmed with Ruminants. The Rehbok (*Pelea capreolus*), Steinbok (*Tragulus rupestris*), Springbok (*Gazella euchore*), Hartebeest (*Alcelaphus caama*), and Koodoo (*Strepsiceros kudu*) were generally seen and could always be found, but now only a few of the smaller " Buck " reward the hunter's toil. It is the scattered Boers who have thus altered this aspect of nature ; they slaughtered the animals for their skins when they found a small price could be obtained for them, and in former days their dress, including boots, were made of buckskin. A buckskin kaross kept them warm or provided the substitute for a carpet, whilst the same animals provided them with a good covering for furniture. No animals could long withstand such persistent slaughter, and to-day the lifeless veld bears witness to one of the Boer influences on nature *. I have often heard old residents and sportsmen describe the panorama of Antelopes once to be seen moving across these scenes, which now are only vast solitudes. It is difficult to estimate the amount of nature's modification through man's influence. Even on these grassy plains, where superficially plant-life looks so poor and uniform, the extirpation of these vast herds of browsing animals must have produced botanical changes and modifications which only a local Darwin could have estimated. But here the growth of trees or shrubs that might have previously been kept down by the ruminants is again frustrated by the periodical grass-fires of the Boers (to be alluded to further on), and thus man again modifies the appearance of nature.

Time passes much more quickly during these long coach journeys than would be expected ; there is a freshness in the air and an absence of restraint that

---

* Methuen, in 1848, describes Springboks migrating in tens of thousands, literally concealing the plains and devouring every green herb, their ravages exceeding those of the locust swarm (' Life in the Wilderness,' p. 59).

Methuen is speaking of the upper regions of the Colony, but the Transvaal must have been equally undisturbed at that time.

contrasts with long railway-trips at home. Thus, though we started at 3 A.M. and did not reach our sleeping-quarters till 7 P.M., fatigue was in an inverse ratio to impatience. Little was seen during this day: a number of widely-scattered Guinea-fowl (*Numida coronata*), which generally frequent more wooded country —" Bushveld "—were passed on the open veld; and occasional Vultures, soaring beneath a cloudless sky, emphasized what has been well called the " Trade-mark of Africa," in the shape of skeletons or carcasses of oxen and horses which had perished by the way and now ornamented at intervals the margins of the road by which we travelled.

We did not start till 6 A.M. on the last day of our route; but the charm of the journey is broken, for we are leaving South-African solitude and approaching the domain of the merchant, the miner, the company promoter, and the speculator; and this combination reaches its apotheosis in Johannesburg, the Chicago of the Transvaal. Long before we reach it clouds of thick brown dust meet and cover us, for a high wind has arisen, and soon the town itself is in view. There is no reason why Johannesburg should not be one of the healthiest spots in the world, its natural position and climate should render it such; defective sanitation a short time back made it a veritable plague-spot, and typhoid fever, often attended with pneumonia that usually attacked both lungs, carried off too many victims, and those who sought gold too often found death. It is the most English town of the Transvaal, and will eventually prove the real capital. In enterprise and business it bears the same relation to Pretoria as the City of London does to Westminster, though both the last and Pretoria are the seats of Parliament. Johannesburg is now [*] in sackcloth and ashes, the occupation of the company promoter is gone, mining companies close almost daily, mining scrip is nearly valueless, and a settled apathy denotes the shareholder. Numbers leave the town, rents fall, the

[*] This applies to the year 1890.

restaurateur no longer reaps a harvest from champagne-drinking customers, and machinery can be bought for almost half its cost in London, with the loss of the heavy transport cost to the Transvaal. But recently the "booms" of Kimberley and Barberton had found a home in Johannesburg, but now it is merely an abode of baffled financiers, unemployed promoters, and more or less ruined shareholders. But Johannesburg will as surely recover from this depression as the French Republic shook off the disaster of Sedan, but it will be only on the ruins of the gambler's wreck with which it is now strewn. The present dreadfully monotonous appearance of the town will be altered when the numerous plantations of trees, which are now growing well, shall have grown more, and perhaps of all towns in the Transvaal, Johannesburg has the future. Even now, in 1891, improvement has commenced, and, as an acquaintance told me in Pretoria, " I can now go to Johannesburg without all my old friends wanting to borrow money of me." Everywhere you are told the same tale by men with whom the times are now hopelessly out of joint—" If I had only realized in time I could have gone home with a fortune." One speculator was pointed out to me who three years back came up from the Cape to Johannesburg with scarcely five shillings; he turned company promoter, and twelve months since could have realized scrip for at least £80,000 (some said £120,000). At the time I saw him he was not worth five pounds. The same thing occurred at the collapse of the "boom" at Barberton. I met a man who had been a canteen keeper there, and who told me he opened a small bar and billiard-room in that town when it was at the height of its pseudo-financial prosperity. As soon as finished he was offered £2000 for it, then £75 per month for four years, both of which proposals he refused. The collapse occurred shortly afterwards, and he sold the place for scarcely the price of the furniture and fittings. He sold, as he told me, "because there was no one who could afford to come in and take a drink."

VIEW IN THE TOWN OF PRETORIA.

We left Johannesburg at 3 P.M., and after a five hours' coach journey reached Pretoria and sought the comforts of Lapin's Fountain Hotel. A railway-line is now being constructed between these towns and the days of this coach-line are numbered.

Pretoria is the seat of government and capital of the Transvaal, and its numerous trees give it a pretty appearance compared with the barren veld on which it stands. It is almost surrounded by high and barren hills and lacks the invigorating climate of the more exposed Johannesburg. The trees which ornament it are not all indigenous and consist principally of a weeping-willow (*Salix gariepina*, Burch.), always a favourite of the Dutch, and here attaining a superb growth; and stately gum-trees (*Eucalyptus*), which either form noble avenues or fringe the borders of the roads. Peach-trees are everywhere abundant, not grown as at home trained to walls, but forming a large and sturdy growth resembling apple-trees. Towards the end of August and beginning of September peach-blossom is so universal as to give a pink hue to the general landscape, and is then one of the most effective botanical sights of Pretoria. This tree, as a general rule, is quite uncared for; it is neither pruned nor manured, though fruit is most abundant but poor in quality: the yellow peach is almost the only kind grown and is moderately hard and flavourless; it is more adapted for cooking, and the Boer farmers use it for making "Peach-brandy," which they sell to the Kafirs. One may obtain an acquired taste for most "liquors," but anything more abominable to a fresh comer than this decoction is difficult to imagine. The peach here seems to revert back to its uncultivated condition, and is found like this in most parts of the Transvaal *. By the 1st

---

* Mr. Wallace remarks that "the peach is unknown in a wild state, unless it is derived from the common almond, on which point there is much difference of opinion among botanists and horticulturists" ('Darwinism,' p. 98).

According to Heyn, this tree "originated in the interior of Asia, beyond even the cherry land, and became known in Italy during the first century of the Roman Empire" ('Wanderings of Plants and Animals from their First Home,' p. 320).

of October the peach-blossom had altogether disappeared and was succeeded by the prodigious bloom of roses, which often constitute whole hedges to fields and gardens. There are a few white blooms, but the majority are of a pale pink colour, mostly single, some semi-double, and there are also small double button-hole blooms which grow in clusters; these roses flower continuously during the whole Transvaalian summer. An occasional passion-flower (*Passiflora*) is also found with the roses and blooms during the same time. Oleanders (*Nerium oleander*) thrive remarkably well in Pretoria. In one private garden are two specimens, each some fifteen to twenty feet high and of the circumference of a large fruit-tree; these at the early part of October became a mass of red bloom and were a glorious contrast to the puny examples we grow in our greenhouses in England. The oleander—cut and trimmed—forms a considerable portion of the hedge which encloses the cemetery. I did not meet with our old friend the Oleander Hawk-Moth (*Chærocampa nerii*), though its non-appearance in my path was probably purely accidental, for I found two other hawk-moths common to our English fauna, which in Pretoria were not scarce and quite unmodified from their usual form: I allude to *Acherontia atropos* and *Protoparce convolvuli*. In gardens the Hibiscus is hardy and blooms freely, but is not so much cultivated as such a handsome plant deserves, whilst the useful and robust "Indian shot-plant" (*Canna indica*) everywhere abounds with its striking foliage and its deep red bloom. Flower-gardens, however, exhibit most of the features of those at home—the geranium, verbena, marigold, stock, dahlia, sunflower, phlox drummondi, and mignonette being very common. Zinnias here attain to particular excellence and growth, and the scattered seed has produced a small wild or degraded form which is found on the hard veld. It will thus be seen that the greater part of the plants and flowers of Pretoria are, like its inhabitants, migrants and colonists. The winter season, during which I arrived with its evergreen and deciduous trees, its orange-trees bearing ripe fruit, and its leafless willows, the August noon and the

March sunrise and sunset, is incongruous in the extreme, and is better described as the cool *dry* season. Towards the end of August gardening operations commence, for the rains are soon expected, and I received a Spring Catalogue of Plants and Seeds from a firm in Port Elizabeth that reminded one of the Carter and Sutton publications at home.

The streets of Pretoria are wide and well designed. Their width, however, had a lowly origin, for they were thus devised and constructed for the convenience of ox-wagons, which could not turn round in narrow roadways. Years hence, when the rail shall have entirely or almost completely replaced the old Boer wagon, this requirement will be forgotten, and those who originally laid out the town will probably be credited with more artistic and less utilitarian tastes. All the Transvaal towns are designed on one scale: given two parallel squares—a church square and market square—connect and approach same with a straight road, and let shorter transverse roads branch off on each side. Pretoria was thus laid out as Pietersburg is to-day, and the grass-grown paths and squares of the last are only like what the first was a few years since. Pretoria is now going through a building phase; its giant government buildings are equal to accommodate the official servants of a State twice the size of the Transvaal; its mercantile buildings are sufficient for twice its present trade, so that business profits have already approached the competitive attenuation. A large market building is being reared upon the market square; the town will shortly be lighted by electricity; churches and chapels abound, and a Church of England Cathedral—small, of course. A water company now supplies pure water—though at a present prohibitive tariff—to supplant the former typhoid beverage of the sluits; there is a permanent race-course, and a prosperous and gigantic distillery sheds a lurid light on three struggling breweries; there are judges, a national flag, and a national anthem—but are these really Boer institutions? and what part have the true Boers taken in producing such results?

BOER WAGON WITH FIREWOOD.

CHAPTER II.

THE BOER.

Where are the Boers?—The Boer a farmer.—Grass-fires and their consequences.—Habits of the farmer.—Peculiar theology of the Boer which governs his life and action.—Boer relations to the Kafirs.—Violence of Church disputes.—President Krüger.—Some causes of the Boer War.—The Boers as soldiers.—Homely life of the President; his great influence with the Boers.—Many farmers now wealthy men.—Physical characteristics of the Boers; their supposed dislike to the British; their mistrust of the Hollanders.

ONE of the first questions I asked after residing a short time in Pretoria, the capital of Boer-land, was, where are the Boers? They are not to be found employed in the Government offices, for here Hollanders are generally engaged; they do not keep stores—at least, so seldom, that the exception proves the rule; there are no Boer clerks in mercantile offices, no handicraft or manufacture carried on by them. British, Dutch, and German are the nationalities which compose the population; but where are the Boers? Beyond the farmers who bring in their produce and firewood for sale, and can be found at the early morning market, the Boer is a visitor at Pretoria, and the same remark applies to all the towns of the Transvaal.

The Boer is a farmer, or, more correctly, a dweller on the veld—he loves solitude and cares nothing for the outside world. I had frequent business relations with one, which occasioned almost weekly visits, and as we became fairly good friends, this farmer may be taken as a typical example of the Boer. This man possesses a tract of 20,000 acres, which is called a farm. Scarcely any of this domain is cultivated; it embraces part of a range of hills which forms a boundary, and contains several isolated eminences as well, whilst in most places its level ground is strewn with rocky débris. These hills are sparsely wooded and it is from them that he obtains the firewood he sells at Pretoria and Johannesburg. He lives in a small and wretchedly kept and furnished house, the most conspicuous articles of which are a small Dutch organ and a large family Bible, for he is a conventionally pious man. He cultivates a very small patch of his farm and leaves the rest, as nature gives it, to grazing purposes, and relies on his flocks and herds. Towards the end of the winter he fires the veld, the withered and dried grasses of which readily burn, and this allows to the new shoots, that will rise after the rains, light and air to commence growth. At that time of the year the illumined horizon almost nightly denotes the process of this primitive farming, and day reveals dismal black areas which tell the same tale. The young grass soon starts, and in a fortnight from the conflagration I have seen scattered and small patches of bright green, even before the rains have commenced. But these continuous fires help to keep the country in its present treeless condition, for nothing but a few stunted trees of the hardest wood can withstand the ravages of the flames, whilst young seedlings have no chance of surviving their first season's growth *. I believe the

---

* The same thing occurred in the early days of the settlers in North America, when the Indians annually burnt the grass on their pasture-grounds. "The oaks bore the annual scorching, at least for a certain time, but if they had been indefinitely continued, they would very probably have been destroyed at last. The soil would have then been much in the prairie condition, and would have needed nothing but grazing for a long succession of years to make the resemblance perfect. That the annual fires alone occasioned the peculiar character of the oak openings, is proved by the fact, that as soon as the Indians had left the country, young trees of many species sprang up and grew luxuriously upon them." See Marsh, quoting from Dwight's Travels ('Man and Nature,' p. 136).

Government have to an extent prohibited these burnings; but as the practice is carried on by the Boers, who are a law unto themselves, the enactment is more honoured in the breach than in the observance.

The Boer farmer usually passes his time in riding about or sitting in his house smoking and drinking coffee. His vrow sees to the house-work, his sons drive the ox-wagon. The living is wretchedly poor and vilely cooked, but the Boer has few wants and is happy if left alone. Kafirs do the farm-work, which is principally attending to the cattle, who neither require food nor water, as the veld provides the first, and they are always kept where some small stream can be found. These people retire to bed at about 7 P.M., but rise early. Illiterate and uneducated to a greater extent than our own rustic population, they possess a keen and intelligent grasp of the government and politics of the Transvaal, and in this respect are intellectually superior to our own men of the shires. They have won their position by hard fighting and hard living. Forty years ago they had to wage war with lions and leopards on their farms, where now scarcely a buck is to be seen, and not only did they struggle against wild beasts, but sustained sanguinary Kafir fights. They showed no mercy to one or the other, but fixed their boundaries and protected their farms. They are the nearest present approach to the old Hebrew patriarchs; like them they value wealth in flocks and herds, and, away from the world in almost lonely wilderness, worship God, and often possess the same strong and unruled passions as were exhibited by some of the biblical personages. Wild tales of wild doings are sometimes told as having occurred in far-away farms; but I incline to the view that these are often exaggerated and that the average Boer is, according to his lights, a citizen pioneer, and a rough, God-fearing, honest, homely, uneducated philistine.

My Boer friend once showed me the two books which appeared to form his library; they were both large Bibles—one in Dutch, which he read; the other in

English, which he did not understand, but which had been *taken as security for a debt*. Both were illustrated in that primitive and almost outrageous fashion which seems to have often inspired biblical artists, and no doubt these pictures considerably influence the minds of these primitive Boers. Science, literature, and criticism being unknown quantities, one can speculate on the theological crudities of these good people. Alone on the veld, with the silent plains often more or less surrounded by the "everlasting hills," the Jehovah of the Jews seems to supplant in their minds the God of the Christian, and these biblical pictures of the pastoral patriarchs must have an attraction and sense meaning to them which are unknown to us. They are alone by themselves and the God of the Illustrated Family Bible. I often ask myself if it is better to be a joyful savage Kafir, or a sombre civilized Boer. The Calvinistic sabbath is supreme on Sunday in Pretoria: the Europeans drive and frequent their hotels and clubs on that day, or, at least, the migratory portion of our race do so; the Boer rejoices in that respectable gloom dear to Scotchmen (at home) and themselves.

To understand the Boer you must understand his theology, which rules his life and guides his actions, and you may as well fight him at once as seek to argue with his prejudices. In the early days of the Boer Trek, they absolutely thought that they would eventually reach Jerusalem*. Their favourite scriptural reading is the Old Testament, and especially the Book of Joshua, where the command to go forward, enjoy the promised land, and smite the heathen was freely adopted by themselves as referring to the Transvaal and the treatment of the Kafirs. It is owing to this feeling that you find towns in the Transvaal called by names such as Bethlehem and Nazareth, and when in their Transvaal advance they approached a river which over-

* For this and some subsequent information, I have the absolute authority of a Protestant clergyman of long experience in the country, whose name I naturally refrain from publishing.

flowed its banks they absolutely thought they had struck a source of the Nile, and called it the Nile River—"Nylstroom," which name it still bears. The same clergyman to whom I have referred also told me that in his travels in the interior he had met most friendly Boers, who told him they could not understand why such an intelligent Englishman should preach to the Kafirs, who possessed no souls. I have been assured by other competent and long residents in the country, that the Boers look upon the Kafirs as the descendants of Cain, and consider any attempt to christianize them as trying to nullify a curse of God. It is difficult to hear these views openly expressed at the present day, and it will be more so in future, now that there is a foreign and critical community around; but it is these esoteric beliefs that often govern the volitions of a people and the government of a country. A friendly Boer once speaking to an acquaintance about Matabele Land, assured him it was a beautiful country and would one day be taken over by the Boers, adding, seriously, "God Almighty never made such a beautiful country for Kafirs."

The Boer treatment of the Kafirs is now certainly much better than it was; but in saying this I feel a great reticence, for there are, and always have been, many Boers of natural kindness of heart, than whom Kafirs could have no better masters. But of others, and in former times, the reverse is the fact, and they treated their Kafir labourers with savage harshness[*]. They had not forgotten the long and sanguinary fights necessary to dispossess the natives of their country, nor of the savage reprisals and murders incidental to the same. Reports are current, for which I will not vouch, that, by degraded Boers, labourers once were sometimes only paid at the expiration of their term and then followed and shot for the recovery of the

---

[*] Burchell gives an instance ('Travels in Interior of South Africa,' vol. ii. p. 95). See also Livingstone ('Popular Account Missionary Travels and Researches,' new edit. p. 28).

money; whilst the poor wretches have often been bound to an apprenticeship of 21 years (which they did not comprehend), any attempts at escape being met with savage floggings and shootings. But these are not purely Boer characteristics. I remember the floggings on English-managed eastern sugar-estates twenty-three years ago, and the flagellations of the Stanley expedition are not yet effaced from memory.

This conflict between Boers and Kafirs still quietly exists. The following was published and guaranteed as true by the 'Uitenhage Times' of this year*:—

"A Dutch farmer and his wife living far north in the Transvaal, with no near neighbours, were surprised one day by twelve strange Kafirs. The farmer, who was outside the house, was bound hand and foot; then, entering the house, the Kafirs began ill-treating the poor woman, but on the suggestion of one of their number, ordered her at once to cook a large pot of mealie pap. This the poor woman did in the presence of the Kafirs, although her clothes were torn from her back, and she was almost naked. When the pap was ready they all squatted round the pot and ordered the woman to get them sugar. She had only a canister and that was in the wagon box; she was told to fetch it. She remembered also at the same time that there was a bottle of poison in the wagon box, which her husband had bought for killing wild animals. Swiftly and secretly she shook the contents of the little bottle among the sugar, and shaking the canister well up, handed it to the Kafirs who helped themselves liberally, with the result that in a short time they were all suffering agonies and went outside one by one. Trembling at what she had done, at the escape she had from death or worse, and for the safety of her husband, the poor creature waited in the house for some time; but eventually went out and found all twelve Kafirs dead, and her husband bound hand and foot in the kraal, but otherwise uninjured. She

---

* Copied by the 'Press,' Pretoria, Feb. 18, 1891.

immediately released him. He quickly buried the bodies of the dead Kafirs, and they resumed their farming-operations as if nothing had happened."

Not only does a crude theology colour the life and guide the political existence of the Boers, but it absolutely threatens to prove the source of their disintegration. From the earliest days of their history church disputes have been readily fomented and violently contested. At the present time one of these is raging to the edification of the whole community, and is consequent to the amalgamation of the two churches, Ned Herv and Ned Gereformeerde, which took place about five or six years ago. There were many dissentients to this amalgamation who refused to join it, and obtained a minister from Holland. It was agreed at the fusion that all properties should be transferred to the amalgamated churches, but this the dissentients refused to ratify, and a lawsuit was commenced in the High Court. But this is little to what occurred at Zeerust last year, when fifty armed Boers entered a Church, took possession of the same, forcibly ejected the minister from the pulpit, and turned the congregation adrift. It is no exaggeration to say that over this dispute the Boers were in measurable distance of civil war. My friend the farmer, of whom I have previously spoken, assured me with anger and sincerity that before any alteration was made with the present government of the Pretoria church, the contents of his rifle would have to be reckoned with, and that a notice would be sent to all the Europeans to avoid the Church Square on a certain day. The President at the time of writing is endeavouring to bring about, if not a reconciliation, at all events some form of arrangement; but feeling runs so high, that a cartoon on the subject just exhibited in a stationer's window was compulsorily removed, in obedience to the threats of angry men, who would otherwise have demolished the windows. The Boer has no sense of humour.

These disputes are a real danger to the State; their

solidarity to the present moment is the only strength of the Boer government, and when once faction commences the liquidators of the present Republic will step in *.

On October 5th occurred the first Dutch church festival during my residence, and of which there are several annually. To attend these the Boers travel in their wagons with their families from all the surrounding districts. Members of the church take the Sacrament, and the younger people are examined and admitted as church members. In former years the Church Square was covered with tents and wagons on these occasions, as the Boer has the right to outspan on the Square, and still possesses the privilege, which does not improve the sanitation of the town. The government now by quiet persuasion endeavours to induce these worshippers to camp outside; but most stand upon their "rights," though, as amongst all people, there are found the few reasonable spirits who listen to advice. I counted thirty-five wagons on the Square this Sunday morning, with the tents under which the families had slept, and towards evening the oxen were gathered together, ready to inspan and start homeward at daylight. Truely these Boers are a strange and unromantic people, a mixture in religion of the old Israelite and the Scotch Covenanter, and a nasty people to manage if their religious prejudices are attacked. I met the President walking to attend this service with his Bible under his arm and his pipe in his mouth. The President, however, belongs to the Dopper branch of the Church, which still remains intact, and the church is opposite the presidential residence, and is regularly attended by his honour, who sometimes conducts the services. The Doppers are the Quakers and Plymouth Brethren of the Dutch Church in the Transvaal. As a rule no instrumental music is used in their services, and no hymns are

---

* Since this was written the President has by conference settled this dispute and has stated "that a serious danger to the State had been happily averted by the combined efforts of the delegates" ('Press' Weekly Edition, Sept. 5, 1891).

allowed, the Psalms of the Old Testament alone being sung. I could not help noticing that the Dopper congregations were better and more neatly dressed, and possessed that appearance of comfortable independence as is observed among the Friends at home.

President Stephanus Johannes Paulus Krüger was born on the 10th October, 1825, in the district of Colesberg in the Cape Colony, and is without doubt the greatest and most representative man that the Boers have yet produced. Uneducated or self-educated, he possesses a very large amount of that natural wisdom so often denied to men of great learning and of literary cultivation. With many prejudices he is fearless, stubborn, and resolute, and he really understands Englishmen little better than they understand him. In his earlier days he has been a somewhat ardent sportsman and a good shot; he has been engaged and honourably mentioned in most of the Kafir fights of his time, and at the end of a rough and stormy life he fills the Presidential chair of a country that has passed Boer aspirations, and attained a financial character and position due to its mineral wealth and the energies of its Colonists, in which Boer industry and Boer influence have played but a small part. Socially he has always lived in a somewhat humble position, and it is to the credit of his nature as a man that he bears not the slightest trace of the *parvenu*. Plain and undistinguished in appearance, he combines the advantages of a prodigious memory with a remarkable aptitude for reading his fellow man, and this last quality would be more valuable were it not leavened by a weakness in resisting flattery and adulation. He is very pious and self-reliant, which is provocative of bigotry and hot temper; and surrounded and approached on all sides by clever and often unscrupulous financiers and speculators, his scutcheon has worn wonderfully well, and his character and reputation passed through many fiery ordeals; he is also a rough diplomatist of no mean rank.

He has been twice married: by his first wife he had

PRESIDENT KRÜGER.

one child; by the second, who still survives, he became the father of sixteen children, seven of whom died when young. Each morning, about 9 A.M., he may be seen driving to the Government Offices, and in the afternoon he holds with pipe and coffee a reception on the "stoop" under the verandah of his house. Here he is daily seen by a heterogeneous assembly of visitors—men with a grievance, applicants for posts, would-be concessionnaires, and even Boers, who seek his advice on family troubles. During the British annexation this was one of the features of Boer life we quite ignored, and for which we afterwards paid a heavy penalty. I have been assured by old and non-official residents, both by English and Dutch, that had "Shepstone remained," the outbreak might have been avoided. The Boers have a patriarchal form of procedure, and when they have, or think they have, a grievance, some elders are deputed to visit Oom Paul, as the President is usually called. President Krüger listens to all they have to say, has a long talk with them, argues the point, hammers in his own convictions with his own *private* reasons and perhaps a few texts of scripture, and the elders go back, explain to their constituents that they now see it all clearly and that they must all be satisfied for the present. Shepstone was trusted and liked, pursued the same policy, possessed their confidence, and had touch with the whole state, and these qualities Lanyon had not.

The longer one lives it is seen that people are governed by their prejudices and illusions, and the mystery becomes greater how the British Colonial Empire exists, governed by pro-consuls who do not possess that idea, and act as though the contrary prevailed. If a West-Indian Governor was transported to rule these Boers, and had the slightest trace of creole blood in his veins, he would be looked upon as a half-bred Kafir, and despised by the most ignorant and unwashed farmer in the territory. The late Boer war should never have broken out. Incapacity caused it; incapacity fought it; incapacity finished it.

The Boers are trained irregular troops from their birth. A lad is first taught to ride a calf, and then a horse. At a certain age he has a rifle given to him, and two cartridges at a time. After a few occasions he must not return empty handed, even if he only brings a bird, or punishment ensues. A Boer in a fight stands behind his horse if in the open, like a dragoon; only the dragoon is taught late in life, and the Boer and horse have grown up together and are one. In a campaign he only requires some dried meat—beltong—attached to his saddle, and a bottle of hollands or water; his rifle and cartridges are secured around his body; his horse will live on the grass of the veld: thus he is fully equipped, and baggage and commissariat unrequired and unknown. Whether in future years they will maintain their wonderful proficiency as marksmen, now that the big game is almost exterminated, is at least open to much doubt; and in after years it is probable that the Boer (not the Hollander), with all his weaknesses, prejudices, and undoubtedly fine qualities, will be but a story that is told. It must always be remembered that not nearly all the Boers were called up in the late war; while some of the richer combatants had two or three young Kafirs behind them with spare rifles, which they loaded and passed to them.

Another cause of the war was the question of the official language. The convention clearly stated that English and Dutch were to be used; but English soon became dominant, and thus a grievance arose. English residents in the Transvaal at the present time must not therefore complain overmuch that Dutch has been made the official tongue.

The President lives in a homely style, and receives no company. His house is not situated in the best part of Pretoria, and there is nothing to denote the abode of the chief of the executive, save a flag-staff and a lounging sentinel. I advisedly use the word lounging, for I passed daily, and have seen these sentinels looking perfect victims of ennui and assuming such positions as would drive a European drill-sergeant to despair.

Thrift marks the Presidential residence. In the spring I witnessed his small front flower-garden being arranged for the coming summer. Two small beds were being bordered by reversed empty glass bottles, the outer border being composed of wine-, the inner border of lemonade-bottles. It was a pity that all the labels had not been washed off; but still the arrangement illustrates the homely and economical, if not artistic, tastes of President Krüger. It must not, however, be imagined that these bottles had been emptied in his establishment, as his honour is practically a total abstainer.

The power and influence of the President are best exemplified at the deliberations of the Raad. When great opposition is manifested to a measure which the President is anxious to pass, he will frequently adjourn the House to the following day, and in the meantime have an interview with the principal dissentients and afford them further reasons for its advocacy. Many of the Boer representatives are bewildered by financial schemes they do not understand, and by political moves which they think affect their rugged independence, and it is then that these personal explanations so largely contribute to the progress of business.

The President is thoroughly in accord with his people as to the belief in direct action of a Special Providence guiding the fortune of the Transvaal; and their present position is still a source of devout wonder to most of the Boers, many of whom really believe that at the last war they actually beat the whole British Army. Of course the President—who has three times visited this country— and the other officials are not under these hallucinations, but looking back at their early experiences much crudity of thought is readily explainable.

An extract from the President's speech at Krugersdorp last year, when the memorial stone was laid of the new Government Buildings there, will readily show his strong feelings on this point*. "Burghers," he exclaimed,

---

* The speeches recently made at the Paardekraal celebration last December more strongly emphasize these views.

"When I cast an eye about this spot, I cannot but acknowledge the hand of the Supreme Being who rules the Universe. The wailing that was once so lamentable here, is now changed into sounds of joy and gladness. From the earliest days the Ruler of the Universe has guided and guarded us. He has protected us from the rude attacks of barbarians when this land saw but few white people—attacks which, if not providentially averted, would have extinguished us. Who does not remember in years gone by the moaning even of women and children when attacked by the savage Kafir hordes? Who can forget when their husbands were fighting for their very lives, their wives brought branches of trees to make a fortification round their wagons and tents? Who, I say, can forget the torture to which some of the *voortrekkers* of this country were exposed?" * Of course the Kafirs would take quite another view of these matters; but like the successful man of business, so it is the victorious people who most often trace the hand of Providence. It seems easy, all the world over, to be thankful for great blessings, and much more difficult to appreciate those more frequent and untoward events which by some have been styled "blessings in disguise."

I have remarked that the President has a slight weakness for adulation, and the following remarks were dished up in the Government Journal on the occasion of his last birthday (1890):—" While he has accomplished much that in other countries would immortalize his name in song and idealize him, as it were, to future generations.... a section led by unscrupulous speculators and mean and degraded journalists, whose name it is not necessary to mention, will probably take no notice of the day at all."

Many of the Boer farmers are now wealthy men—not only in land but in cash—owing to the large sums paid them by mining companies and syndicates for their auriferous farms. Frequently this cash is kept in some

---

* As reported in the 'Press' of Pretoria, Sept. 20, 1890.

sure hiding-place, for the Boer has not yet acquired a confidence in Banks; and I have been assured, on good authority, that some of these primitive folk who have deposited sums at Banking institutions have called and asked to see their money. A ready cashier will at once produce a quantity of gold from his drawer, and confidence is restored. In former days little cash was handled by the Boers; they possessed large farms or, rather, unworked tracts of land, but money was scarce, and heavy and laborious wagon transport was undertaken for small sums.

In stature the Boer is tall and strongly built, but seldom stout. Living in one of the most healthy and invigorating climes—I speak of the high veld—he possesses, as a rule, a splendid constitution and a capacity for much more work than he cares to undertake; his ordinary spare and meagre diet prevents much aptitude for corpulency. For bathing he has no desire, and he is as economical in the use of soap as any white race found on the globe.

It is generally thought, and especially in this country, that the Boers have a hatred for Englishmen. This is a fallacy, for, in spite of all that may be said and done, the Britisher is respected though not loved. His word is taken, his honesty accepted, but his arrogance is overestimated. The Hollander, on the contrary, though so near by blood, is neither respected nor loved. Englishmen improve the country, even if their old colonial instincts prompt a desire to fly the old flag; the Hollander is often a financial parasite. The Englishman will toil if he can reap: the Hollander will reap if possible, but not toil. It is the Hollanders in the Transvaal who dislike the English, and are alike detested by the Boers.

The longer I remained in the country, the more absurd it appeared for the English to have lost it. England could have worked well with the Boers by proper management, and Hollanders would no longer have had the opportunity of exploiting them. But a Boer is a plain man: he can understand an English farmer but not an English aristocrat, and why a pious straight man like Sir Bartle Frere could not manage the President is only explainable on faults of individuality and not of character. An Indian proconsul, with his acquired hauteur and social exclusiveness, which are so often more apparent than real, is no diplomatist for the Transvaal; the imported Hollander, however, is, and has been, too often a financial curse to the Republic.

The Boer to-day is what may be called "smart" in the little business he does with the community, and this applies principally in the relation of sample to bulk of the produce he disposes on the market. But it must not be forgotten that he has learned much of this through bitter experience and from those who now speak the strongest on the subject. The Chosen People swarm in the Transvaal and have pitted their financial and commercial talents against the once unsophisticated farmer, with of course one result. One Israelite, whom I frequently saw in Pretoria, and of whom many good and other stories were told, had acted as produce agent for a Boer, whom he generally cheated of a few pounds, in the settlement. One day the Boer arrived indignant, and with a "ready reckoner" in his hand demanded a balance. "What book have you there?" enquired the clever Semite. "A ready reckoner." "Let me see it;" and then returning it contemptuously to the dissatisfied one, added with withering scorn, "why it is last year's edition you have got!" The Boer retired mystified.

The Boers seldom laugh, and have no gaiety; they know neither the pleasures of music, literature, nor even the table; they are fond of shooting, and are perhaps the finest shots in the world, though they have now nearly exterminated all the big game. No people have ever made the wagon such a home, or driven it with such skill. They possess all the virtues of home life, and are sober and thrifty, drinking perhaps less alcohol and smoking more tobacco than any other people. They have a character for inhospitality, as many a lone and weary traveller or prospector who has sought the shelter of their houses, or asked for food, will declare. But the Boer wished to be left alone, his early treks were made for solitude as well as freedom. He is amazed at the developments of the big towns and prefers the quietude of his farm. Other people are now supplanting him in the Republic; his habits of retirement will prove his effacement, and his want of education will consign him to oblivion unless he treks still further on. If the records of these early treks could only be gathered before the chief actors, who are now old, have passed away, much zoological and ethnological information would be saved, often of no mean importance; whilst deeds of endurance and heroism would be recorded, and a love of exploration disclosed, that would rival the doings of some of our modern travellers who write big books and receive great receptions.

I have sought to be impartial to the Boer, whom I respect but cannot love; and my principal remarks apply to the real Boers, the farmers, the dwellers on the plains, and not to the official Krügers, Jouberts, Smits, and others, who really constitute the Boer aristocracy, and no more represent the average population than the inhabitants of the West-end of London are typical of the real population of England. The shadows are deepening over these hardy farmers, the pen will conquer what the sword could not subdue, and they must be either absorbed in or fly from the busy mercantile population that is now surrounding and must ultimately dispossess them. In the nineteenth century there seems no room for this old pastoral life, especially when nature has baited the soil with auriferous deposits; but I shall ever remember the family wagon of the Boer when my fancy recalls the peaceful wilderness of the veld.

(38)

Chera progne. Male in nuptial plumage.

CHAPTER III.

PHASES OF NATURE AROUND PRETORIA.

Natural aspects in the dry winter season.—Orthoptera and Coleoptera.—Commencement of the rainy season.—Protective resemblance in butterflies.—Vegetable tanning-products.—Survival of spined and hard-wooded trees in the struggle for existence with herbivorous fauna.—Baboons.—Bad roads.—A Boer farm.—Grass-fires.—Dust-storm.—Vast quantities of beetles under stones.—Bad weather and heavy losses in live stock.—Appearance of winged Termites.—Swollen streams and their dangers.—Accidental dangers in animal life.—Birds of Prey.

To a naturalist who has seen the glorious profusion

of plant and animal life in the Eastern tropics, the bare, withered, treeless veld as it appears in the neighbourhood of Johannesburg and Pretoria at the end of the dry or winter season is most dispiriting. The few thorny acacias are almost universally destitute of leaves, the few plants that should be green are more or less covered with fine brown dust, and the only charm is the clear and invigorating air and the bright blue sky. Insect life is almost absent at this period. The first insect I saw was a large locust with red underwings, flying along a road in Pretoria, and chased by dogs who eventually secured it—the strangest hunt I ever witnessed. At this period, the end of July, five butterflies alone enlivened the scene — the ubiquitous *Danais chrysippus* was the most prevalent, a close ally to our English Clouded Yellow was found in *Colias electra* with its two forms of the female sex, a small Teriad (*Terias brigitta*), the wide-ranging *Pieris mesentina*, and last, but not least, an old friend, known in England as the " Painted Lady " (*Pyrameis cardui*) *.

A few orthopterous insects are even then found amongst the dried and scanty herbage of a cast-iron soil; but these are few in number, still fewer in species, and poor in size and colour. The coleopterist now only finds his prey under stones near banks of streams or in other damp places, and it was in such spots that I secured rare species of *Chlænius, Tetragonoderus*, and other good things, besides finding the large earwig (*Lapidura riparia*) sometimes seen in the south of England. Even these were, however, very scarce, and the searcher for Carabidæ must have energy, patience, and experience. The stones must rest in spots neither too dry nor too damp for these small and usually brilliant beetles to seek a shelter beneath them, whilst the labour of turning over the numbers under which nothing is found becomes monotonous and fatiguing. On the dry and hard ground of the more open veld, the removal of a large stone or piece of rock frequently dis-

---

* Excluding the Teriad, these butterflies are found all the year round.

closes the retreating form of a small and elegant beetle down a narrow hole made in the irony soil. It is under these stones that vast colonies of ants are frequently found, and in the immediate neighbourhood of these it seems useless to search for beetles, save the small *Pentaplatarthrus natalensis*, which, as well known, is a messmate of the ants. Two species of "Bombardier Beetles" are not uncommon; one of these, *Pheropsophus litigiosus*, is found under and amongst stones by the banks of streams. When handled its peculiar and protective anal explosion gave a distinct sound, and a considerable puff of smoke was emitted *; the resultant excretion thereby not only deeply stained my fingers, but actually in one case caused a feeling of a smart burn which lasted for fully a minute. The stain on my fingers was indelible for five days.

One naturally became anxious for the promised rains, which would transform this sterile scene, and afford some illustration of African insect-life. On August 5th the clouds gathered about 4 P.M., and a strong wind arose bringing clouds of dust from Pretoria, and a moderate shower of rain. But this was of short duration, and in half an hour the wind blew strongly from the opposite quarter and carried the dust back again. This was premature rain, and no more denoted the arrival of the wet season than a warm January day in England is a harbinger of the spring. But in August the nights became warmer, trees commenced budding, and in a few places the veld showed signs of fresh life. In some spots a few more butterflies now appeared. *Junonia cebrene* and *Hamanumida dædalus* took wing, and the last named afforded me an opportunity of observation which supplemented, if not corrected, some previous statements. Since Darwin taught naturalists to seek and read the story of cause and effect, where genera and species had alone been formerly observed, butterflies have been much studied

---

* It is possible for these Bombardier Beetles to have their artillery artificially discharged after death, as I once found on pinning some dead specimens.

and with great effect on the questions of "mimicry" and "protective resemblance." It has been eloquently remarked by Mr. Bates, that on their wings "nature writes as on a tablet the story of the modifications of species, so truly do all changes of the organization register themselves thereon"*, and a cabinet of butterflies in the possession of a competent naturalist now not only exhibits what used simply to be called the "works of nature," but absolutely in many cases shows how nature works. *Hamanumida dædalus*, formerly and generally quoted by its better-known synonym *Aterica meleagris*, has been recorded as a good instance of "protective resemblance." Mr. Wallace has recently stated that it "always settles on the ground with closed wings, which so resemble the soil of the district that it can with difficulty be seen, and the colour varies with the soil in different localities. Thus, specimens from Senegambia were dull brown, the soil being reddish sand and iron-clay; those from Calabar and Cameroons were light brown with numerous small white spots, the soil of those countries being light brown clay with small quartz pebbles; whilst in other localities where the colours of the soil were more varied, the colours of the butterfly varied also. Here we have variation in a single species, which has become specialized in certain areas to harmonize with the colour of the soil †. But in the Transvaal this butterfly never settles on the ground with closed wings, and the only example sent from Durban by Colonel Bowker to Mr. Trimen was described as "settled on a footpath with wings expanded" ‡. I saw and captured a large number of specimens, and always found them resting with wings expanded, and nearly always on greyish-coloured rocks or slaty-hued paths, with which the colour of the upper surface of their wings wonderfully assimilated. Large tracts of bare ground of a reddish-brown colour exist with which the under surface of the wings would be in perfect

---

* 'The Naturalist on the Amazons.'
† 'Darwinism,' p. 207.
‡ 'South African Butterflies,' vol. i. p. 310.

unison; but though I watched for months to see a specimen thus situated, and with its wings vertically closed, I never succeeded in doing so.

Thus, if the reports as to its habits in Senegambia, Calabar, and Cameroons are correct, we have not only a change of habit with difference of latitude, but also what I have elsewhere ventured to describe as an instance of "Compound Protective Resemblance"[*]. For we see that while in Senegambia, Calabar, and Cameroons, where (according to report) the butterfly always settles with wings vertically closed, and which "so closely resemble the soil of the district, that it can with difficulty be seen, and the colour varies with the soil in different localities," in the Transvaal and Natal it rests with horizontally-expanded wings[†], by which its protection is almost equally insured by the assimilative colour of the same to the rocks and paths on which it is usually found. My friend Mr. Trimen, with whom I discussed this matter, suggested that I should observe whether the upperside might be protective in the wet season, and the underside in the dry; but whatever may be the case elsewhere, I saw that its habits were uniform in the Transvaal in both the dry and wet seasons.

I was afforded a good opportunity of watching the gradual approach of spring and summer, with their transforming effects in the production of plant and insect life, as business weekly compelled me to drive some 15 miles out from Pretoria to a Boer farm, on the hills of which grew a tree capable of supplying bark for tanning-purposes. This was called the "sugar-tree;" but the bark was coarse and possessed little strength. The best and strongest tanning-material in the Transvaal appears to be the leaf of a tree (*Colpoon compressum*)[‡],

---

[*] 'Nature,' vol. xlii. p. 390.

[†] Although as a general rule the species of Nymphalidæ, to which family this butterfly belongs, do rest with vertically closed wings, the species of the tropical American genus *Ageronia* have a similar habit to *H. dædalus* as observed in the Transvaal.

[‡] For the exact identification of this species, I am indebted to the Curator of the Durban Botanic Gardens.

called by the Boers "Berg bas." It is found scattered about in the woody portions of the country, but grows most plentifully—at least it was there from whence we obtained our largest supplies—on the hills of the Waterberg district. The sugar-bark was obtained on the farm I have mentioned, which was situated in what was known as "Ward Crocodile River," and at no equal distance from Pretoria could a greater diversity of scenery be found. I drove in a "spider" drawn by what appeared to be two sorry nags; but in this country it is such looking animals which show an endurance and aptitude for the peculiar "roads," not to be equalled by better horses at home. The first part of the journey was along the somewhat good road which crosses the level veld towards the Crocodile River; but after an hour's drive we turned off, and leaving the plain, struck across country for the mountains or kopjes on the left. At this spot, on a clear day, these ranges could be seen rising one above another in the distance, the farthest only seen in greyish outline, and a blue sky and fresh air prompted that joyous feeling that mountain slopes produce under similar circumstances in all parts of the world. The shadows of these bare hills are thrown one upon another in an almost artificial manner, sometimes in colour nearly black, and in shape frequently an almost perfect parallelogram, as though the slopes were a screen on which a solar lantern threw its magic shapes. The road now becomes much worse, and large rocky stones are freely strewn about the track over which we drive. Trees are more plentiful, but are principally long-spined acacias and "iron" and other hard-wooded species. These trees are the silent witnesses of what was once the head-quarters of the ruminant mammalia, now practically exterminated or driven back by the incessant warfare waged against them by the Boer farmers, and by the opening up of the country to a mining and mercantile civilization. There was a time when a deadly struggle went on between the plants and trees of this region and the vast herds of herbivorous animals that swarmed over it.

These long-spined acacias and hard-wooded trees alone possessed an adequate resistance to such attacks, and their survival proclaims that they were the fittest in the long struggle for existence which in that phase has now passed away. To-day their danger is from the grass-fires of the Boer or in their capacity for supplying fuel.

We meet the river—which in its serpentine course has twice to be crossed—the first time at the base of a quartzite cliff which affords a dwelling-place for a small colony of baboons*, one of which, that has been late in returning from his nightly excursions, I have sometimes surprised early in the day. It was at this spot also where one could meet and secure a specimen of the migrant "European Bee-eater" (*Merops apiaster*), a bird of the gayest plumage to be found in the neighbourhood; whilst it was here and beyond my reach that I have watched the wild and majestic flight of a *Charaxes* butterfly, a species I was never able to secure. This river, so clear and shallow during the dry season, was sometimes found impassable during the rains. Our way becomes more tortuous as we ascend and descend the ridges of the higher ground till we reach about the roughest piece of road that man ever drove over, or that can be surpassed in South-African driving. A hill with a surface of broken rock and bearing a few trees has to be crossed; the road, if it can be called one, rises steeply up one side, crosses the crest, and abruptly descends the other extremity. The whole way is one mass of broken quartzite jumbled together in titanic undulation, and one hardly knew at which to be most thankful—for having driven up one side, or safely travelled down the other. A narrow road ensues, with trees overhead, a river beneath on one side, and the quartzite hills rising high and rough-hewn around us. Great blocks of rock strewn here and there, now peacefully surrounded by herbage, tell the story of the wild crash in which at some bygone time, they have broken away from the parent block above and plunged head-

---

* These kill the young sheep, and are therefore assiduously shot by the farmers.

long. It was on these rocky cliffs that I made my first acquaintance with the *Euphorbiæ* and *Aloes*, so typical of the South-African flora.

The farm is soon reached in all its simplicity. Twenty thousand acres, including hills, is not a bad stretch of country for one man to own; and when it is considered that nearly the whole of this tract is in the same condition as it was when first allotted at the time of the early Boer settlement, with the exception that all the large game, including lions and leopards, are now slaughtered or driven back, a peculiar feature of the Transvaal problem is apparent. Sitting on one of the hills which surround this homestead, and looking at the lonely grandeur of the scene, one wonders why these Boers, under the laws of the average of genius, have not produced a Robert Burns or the founder of some new religion. It was on these hills that our Kafirs felled the trees and stripped the bark, and looked forward to my weekly visit with their wages, as "one day further on" their return to their kraal with the cash sufficient to negotiate the arrangement for another wife.

Towards the end of August the nights became decidedly warmer, though no rain fell. Dragonflies somewhat suddenly appeared hovering over small ponds, of which *Crocothemis erythræa* and the giant *Anax mauricianus* were the most common, two oak trees growing near the Church Square were approaching fair leaf, and the universal peach-bloom gave a warm colour to the whole scene. Small patches of *Sedum*, sp.?, were blooming on the adamantine veld, and the representatives of butterfly life were increased by the appearance of some species of *Acræa* and of *Papilio demoleus*. A few bugs (*Lygæidæ*) could now be obtained by sweeping; but the rains were still absent, and the full spring was not yet, though small water-beetles (*Aulonogyrus abdominalis*) in the noonday sun skimmed the surface of the clear brooks, on the shady banks of which quantities of the Maidenhair Fern (*Adiantum*, sp.) were now growing luxuriantly. During this dry cool season of the year many strange insects are found in a semi-torpid condition under stones. In these situations I have found *Carabidæ, Staphylinidæ, Paussidæ, Curculionidæ, Chrysomelidæ, Gallerucidæ*, and *Coccinellidæ* among beetles, and *Pentatomidæ* and *Pyrrhocoridæ* among Hemiptera, but few in numbers, and at this season the Pretorian province of the Transvaal is most uninviting to the entomologist. The weather is still like spring at home, the nights and mornings quite cold, and it is difficult to believe that one is living in Southern Africa.

With the advent of September the thorny acacias were found to be thickly covering with leaves, and the long white thorns being thus hidden, their striking protection was scarcely visible. It is only when these trees are bare of leaf that it can be clearly appreciated what impregnable objects they are to any herbivorous animal. The grass-fires were now being pushed on by the Boers, and I frequently noticed that blackened areas of some miles in extent, often embracing several hills, replaced what quite recently resembled in colour a field of ripe oats. The veld has thus three aspects—the dull ochraceous hue of the dry season, the blackened tint following the spring fires, and the green coloration of the summer. Numbers of insects in their immature stages, as well as small reptiles, must be destroyed by these fires, and, as remarked before, small seedling trees have little chance of reaching that stage of growth and hardihood necessary to survive the conflagration.

During September some of the acacias bloomed, consequent upon the undoubtedly higher temperature, and these flowers were visited by swarms of Diptera; but still scarcely a beetle was to be seen, excepting a few *Scarabæidæ*. The butterfly list was increased by *Hypanis ilithyia, Precis cloantha*, and *Catopsilia florella*, whilst *Anoplocnemis curvipes*, on the wing, gave promise that Hemiptera would soon be seen, though representatives of various families of this order, as well as of Coleoptera, could still be found, but *only* under stones. On Sept. 25th a heavy shower at midnight gave hopes of the advent of the rains; but it did not last long, and in the morning scarcely a sign of wet was to be seen. It was not till October 4th that the rainy season really commenced. All day the weather had been close and oppressive, and those who suffered from weak chests had found much inconvenience. In the afternoon occurred our first Dust-storm, and that of unusual severity. No rain had fallen for five months, and the consequent accumulation of dust in the town and on the neighbouring roads can be easily imagined. It was under these circumstances that a heavy south-westerly gale broke upon us, and a vast and majestic cloud of tons of dust and small stones rose high in the air, and rapidly reaching the centre of Pretoria, soon cleared the streets both of passenger and vehicular traffic*. Rain fell for about an hour, vivid lightning subsequently illumined a particularly dark night, and nature proclaimed that the long-continued drought was broken up.

It was on the day following this storm that I visited some rocky débris lying under an acacia-tree on the open veld. To my surprise I found under these stones thousands of two small species of beetles (*Rutelidæ*) belonging to the genus *Adoretus* (*A. luteipes* and an unidentified sp.), in a perfectly dormant condition, though the light and warmth of the sun soon aroused them, and they made for fresh shelter. Three weeks previously, and again a week later, I examined this spot and turned over the same stones, but the beetles were on both of these occasions only represented by a few specimens, and not by the prodigious quantities which I have described. It appears that this vast aggregation was preparatory to their segregation and dispersal over the surrounding area, as subsequently during the evenings these *Adoreti*, like moths, flew into rooms, attracted by the light.

Showers of rain fell on the days that immediately succeeded the storm, and the wind shifting to the east blew a gale and was bitterly cold. In the evening the

---

* In Johannesburg houses were unroofed.

streets were nearly empty—a fire indoors would have
been comfortable, and a heavy ulster was found none too
warm. At night thunder rolled, and the rain falling
with a rattle on our roof of corrugated iron effectually
banished sleep. In a few days reports came in from all
sides of the Transvaal detailing the severity of the
weather. From Barberton we learned that the intense
heat prevailing there for some time had broken up, and
a furious gale had ensued, followed by heavy rain and
intense cold, the surrounding mountains being snow-
capped. From Ermelo news came of a heavy snow-storm
and anticipations of severe losses in live stock. In the
Klip River country the snow also fell, and one farmer
lost four hundred sheep and twenty horses within
twenty-four hours. At Lydenburg snow fell in some
instances two inches deep, though this weather was
pronounced to be an unusual phenomenon. Between
Pretoria and Barberton, on the high veld, I was assured
that thousands of sheep and oxen were lying killed by
the cold acting on their present half-fed and poor condition. All the month of October was wet and usually
cold; the veld had become perfectly saturated, and we
now only anticipated a clear sky to enable the increasing
strength of the solar rays to act as the magician's wand
in the transformation scene of Nature.

During one of these rainy October days the air was
filled about noon with numbers of a small winged form
of the Termite, or White Ant (*Termes*, sp.), which
pursued a slow flight through the drenching rain. I
found them emerging in continuous columns through
small holes on the level veld, which scarcely allowed for
the passage of more than one, or at most two, at a time,
when they immediately took wing and hovered around.
They were, however, being devoured by the large and
handsome frog (*Rana adspersa*), which I had not seen
before, and which also issued from holes on the veld.
These frogs stationed themselves near the holes from
which the termites emerged, and literally gorged themselves to repletion. A smaller and duller-coloured toad
(*Bufo regularis*) and a handsome green and spotted frog

also assisted at the banquet. The termites began to issue about noon, and were still flying, though in less numbers, at sunset; but none were seen the following morning, and the toads and frogs had likewise disappeared, though it was still cloudy and wet. I caught many of these termites, but, though I put them in a strong cyanide bottle at once, they almost

Bufo regularis.   *Rana adspersa.*
BATRACHIANS DEVOURING TERMITES.

invariably dispossessed themselves of their wings before death. About a month afterwards (in November) a much larger species (*Termes angustatus*) as suddenly and in equal quantity appeared. This time they were largely destroyed by the Cape Wagtails (*Motacilla capensis*), which, however, fortunately for the termites, were in far less numbers than formerly, as they appeared to have left for their breeding-grounds. A small terrier dog in our possession also played havoc with the ants, which it not only caught, but eat in large numbers.

After a fortnight's intermittent rain, the weather became sufficiently favourable, or rather the roads were once more passable, for another visit to my Kafirs at the Bark farm. A new world of animal life now met the view as I drove along the roads, which in many places were composed of marshy mud, where on my last visit I raised clouds of dust. In Coleoptera giant Anthias (*Anthia thoracica* and *A. maxillosa*) were seen foraging about, and the huge *Manticora tuberculata* was very abundant, whilst *Polyhirma macilenta* ran about the roads where the surface was sandy and gritty. In this way I frequently stopped and obtained some fine species. In the wooded tracts I found Cetonias on the wing, many adhering to the leaves of trees, and one (*Diplognatha hebræa*) even on the long stalks of last season's dried but now damp grasses. In the wet but scant herbage Blue Cranes (*Anthropoides paradisea*), usually in pairs, searched for the orthopterous insects which now almost daily became more plentiful, whilst the Widow-bird (*Chera progne*) had now again developed its long tail-feathers for the breeding-season, and frequented the long sedgy grasses that grew on the marshy portions of the veld. These long tail-feathers appear to offer a direct hindrance to flight, and the birds always seemed to proceed with difficulty and great encumbrance, like a Court Lady dragging a heavy train.

Nature frequently reminds mankind of her forces, and she did so with these heavy rains: small spruits became torrents, and insignificant rivers raging floods. As usual, accounts slowly came into Pretoria—for it is the press which allows civilized man to rise above tradition and hearsay, and newspapers give to prosaic life the romance of current history. The "Six-mile Spruit," a stream through which the coaches drive, and at a distance from Pretoria which its name specifies, came down with a suddenness that has made it famous among the streams

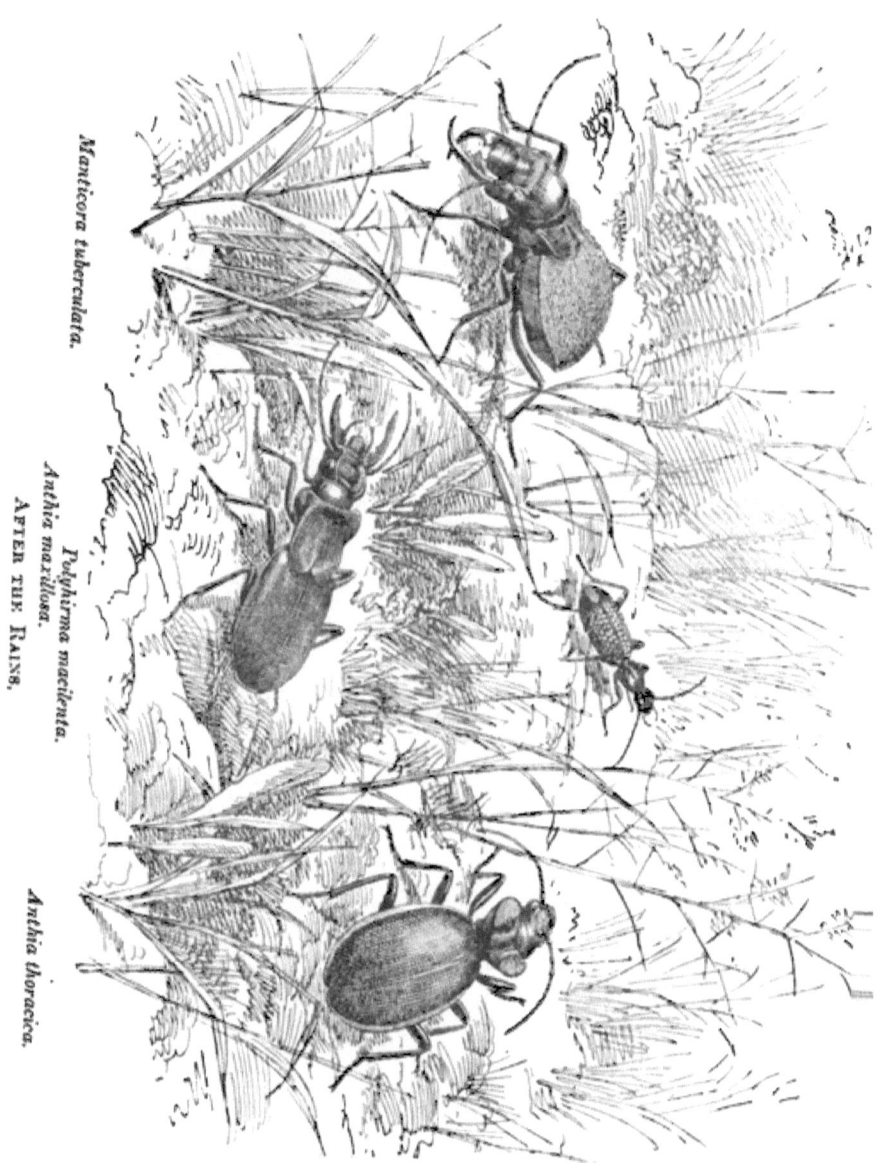

Manticora tuberculata. Polyphirma macilenta. Anthia maxillosa. Anthia thoracica.
AFTER THE RAIN.

of the Transvaal. It came down with such force, with a quantity of water so enormous, and so swiftly, that upwards of a hundred oxen that were feeding on the banks of the river were swept away and drowned. Carcasses, which were estimated at about 150, were found when the waters subsided, either washed out on the banks or stuck in trees at the turns of the river, so that the Kafirs and vultures had an opportunity for high banquet. But besides oxen numerous Kafirs, mostly cattle-herds, were swept away by the flood; bodies were seen floating down the stream, and others were found on the banks. A buck-wagon with a span of eight mules and two horses arrived at the spruit towards evening. No sooner did the wagon reach the middle of the stream than it was completely turned upside down and swept away, and the bodies of the mules and horses were found the following day entangled in the harness. A heavy hailstorm passed along the valley of the river, and the hail floated down in such quantity that large blocks of ice, several feet in thickness, were carried down the stream.

One day, in the early part of November, I was able to appreciate the sudden rising of these Transvaalian streams. Behind our works, and crossing the veld, was a narrow deeply waterworn river-bed, at the bottom of which usually flowed a shallow streamlet of not more than a few inches deep, and which I easily strode across in the morning. By 3 P.M. this was a roaring rushing current some ten feet deep. This was caused by two very heavy falls of rain, the first one continuing from about 12 to 12.45 P.M., the second and heaviest lasting from about 2 to 3 P.M. The roads were flooded, the water poured down the sides of the hills in streams, and a roaring noise could be heard some distance from what was only a few hours previously a shallow brook. The water, which was of a deep reddish clayey hue, boiled, whirled, and tore down its bed, the many bends of which caused whirlpools and some nasty backwaters. Nowhere wider than 20 feet, and generally narrower, it would have been certain death to have fallen therein. By

4 P.M. the principal rush of water had drained down, and, though the river was still full, it was now silent, and by the next morning it had almost resumed its ordinary obscurity. Thus sudden and dangerous are the results of these heavy rains. Insects in numbers must have been carried away, and some were found in a wet and exhausted condition clinging to low shrubs, and I thus obtained an orthopterous insect (*Pycnodictya adustum*) and the rare Dragonfly (*Tramea basilaris*), neither of which I ever found again. It was interesting to observe the different sculpture in many parts of the banks after this visitation, and one could now understand how it was that the usually shallow brook flowed at the bottom of so deeply an excavated river-bed. As November advanced flowers and insects became more plentiful, and the most abundant beetle was the large heteromerous red-striped *Psammodes striatus*. These beetles, when they first appeared, were most abundant on the roads which crossed the veld, and, though globular and ungainly in shape, yet actively ran on their high legs, but were so numerous that we crushed many under the wheels and horses' hoofs as we drove along. I believe that these form a considerable portion of the prey of the different species of *Anthia*, and also of the *Manticora*, which actively patrol these spots; and in the dry season I had often been puzzled to explain the number of empty shells of the *Psammodes* which I found strewn about. Beautifully marked Longicorn beetles enlivened the scene, and about this time I was much struck with the numbers of two species of Weevils (*Polyclaeis equestris* and *P. vinereis*), that literally covered the acacia and other shrubs to be found on the veld. These two species were always found together, and it was only because the sexes of each could be found, and often *in cop.*, that my doubts as to their being one species were dispelled.

When we first arrived and saw the long white spines of the acacias, I involuntarily wondered why no signs were seen of the larder of the Shrikes, of which there are a fair number of species in the Transvaal. I at length came upon their haunts, and, strange to say, a frog was the first animal seen impaled. I afterwards found that small lizards were very common victims, and the black-and-white shrike (*Lanius collaris*), the most abundant species in our neighbourhood, was as fearless as it was predatory. I once followed one of these birds amongst some trees to see what it held in its beak, and approached close to the shrike before it took flight, when, after impaling a large mole-cricket close before my eyes, it flew away to another tree in the vicinity. But nature is "red in tooth and claw"; the small clump of shrubs that bore these impaled lizards were visited by numbers of the previously mentioned weevils, many of which fell victims to the numerous spiders that inhabited cocoon-like structures and spread their webs across the ends of the small branches. Accidents also happen to all living things alike. I once saw a weevil (*Polyclaeis cinereis*), when suddenly alighting from flight on the stems of an acacia, run a spine through one of its underwings and hang suspended. I liberated this unfortunate after watching its ineffectual struggles for some time, and if it had eventually extricated itself from the thorn, it could only have done so at the expense of a mutilated wing. On a subsequent occasion I saw a migratory locust strike in its flight the barbed wire used in fencing, and impale itself by driving a spike through the front part of its head. These untoward events occur much more frequently than we suppose; man has not a monopoly of the miseries of life. Amongst the Vertebrata, if the sportsman or naturalist examined the skeletons of his prizes, he would occasionally find the traces of past fractures and dislocations; and even amongst insects this can be discerned, but usually, or most clearly, in the large Orthoptera, whose long limbs are particularly liable to the accidents of field and flood, and the size of which renders the marks of these misadventures more visible than is the case among smaller insects.

Many birds of prey visit the immediate neighbourhood of Pretoria, and are a considerable danger to young poultry reared in open situations. The most common of these depredators is the Rufous Buzzard (*Buteo desertorum*) and an occasional Yellow-billed Kite (*Milvus ægyptius*). These birds, especially the former, were par-

Buteo desertorum.   Post of Observation.

ticularly numerous about the month of December, and were a great source of trouble to the small squatters on the veld, who erect their shanties (for no other word will adequately express the poverty of these dwellings) on the outskirts of the town. Most of these people kept a few poultry, and their young chickens and ducklings too often served as food for the active and rapacious birds. I skinned several specimens that were shot about this time, and they were lined with layers of yellow fat, similar to what is found in an overfed Christmas goose. These buzzards were particularly fond of sitting on the telegraph-poles that crossed the veld, or using the tops of ant-hills as a post of observation, and were a terror to all the domestic birds of the neighbourhood. The dread of impending evil sits as heavily on the minds of these ducks and fowls as the fear of poverty chills the heart of so many men; and I once witnessed this instinctive or inherited terror, in the wild alarm shown by a brood of young ducklings at the shadow and sudden appearance of a tame pigeon just above them. This poor pigeon unwittingly caused a Buzzard panic, and proved unmistakably the frequency of a real danger, though giving at the time a false alarm.

DENDRITIC MARKINGS IN QUARTZITE.

CHAPTER IV.

PHASES OF NATURE AROUND PRETORIA (*continued*).

Geological features.—Dendritic markings.—The highlands and the sea.—Heavy rains and floods.—A protected butterfly and its enemy.—Mimicry.—Cicadas.—Species found both in England and the Transvaal.—The Secretary-bird.—Vultures.—Locust-swarm.—The Paauw and other Bustards.—The Monitor.—Partridges.—Evolution and struggle for existence.

THE geological feature of the country surrounding Pretoria is quartzite, through which the granites frequently outcrop, as may be best observed by following some of the spruits and smaller watercourses. This quartzite also largely contributes to the rocky mass of the Magaliesberg mountains, which form so considerable a shelter to Pretoria, as I had a good opportunity of observing during the blasting-operations by which a carriage road was made through the rocky defile called the Daas Poort. Dynamite was the agent used to rend the stratified quartzite, and in the blocks thus detached and broken up dendritic or arborescent markings abounded.

These results of the infiltration of oxide of manganese so strikingly resemble the impressions of ferns as to make one believe, on seeing them for the first time, that veritable fossils had been found. On one afternoon, whilst entomologizing in a river-bed just beneath the field of these operations and unaware that a number of mines were just ready for explosion, we were only observed and warned just in time to enable us to retreat in a shower of small rocky débris, and thus to fulfil the parts of spectators and not victims.

Some surface auriferous deposits may be found around Pretoria; but these extend to no depth, and gold is practically absent as a mining industry, though thirty-five miles south is found the celebrated *Main reef* which has created Johannesburg. Pretoria must, in a mining sense, rely on its argentiferous copper and lead, with which is also found antimony. No observer who stands upon or looks at the mass of the Transvaalian quartzose matter can help speculating on its origin. That it was due to the erosion and disintegration of some former vast accumulation of granitic rocks is plain geological interpretation; but where were these granitic masses situate?*

There is a charm in life on this high tableland six thousand feet above the sea and which really forms the heart of the Transvaal; but to all it does not convey the same impression. A recent lady traveller has remarked, "to me it seems quite natural that the centre of a continent is its healthiest point, for one is furthest away from the detestable moisture of our vaunted sea-breezes. Of course we praise sea and sea-breezes here because we cannot get away from them "†. But clever sayings are not always of universal application. It is easy to understand the physical basis of thought, and how a particular constitution may be vigorous in the Carpathians and depressed by the sea; but in the Transvaal the recurrent hills and plains of the tableland only seem to accentuate

* Mr. C. J. Alford has recently remarked:—"Certainly the land-surface from which these materials are derived has long ages ago been obliterated from the surface of the earth." ('Geological Features of the Transvaal,' p. 14).
† Miss Dowie, British Association, 1890.

the loss of the sea that we have left behind. I have frequently driven over the grandest undulating scenery in the most clear and faultless weather; but the feeling always was that behind yonder headland must be the sea. A long residence in England impresses its particular features of physiography upon the mind, and I found I was apt to read nature with a similar insular bias as that with which one studies foreign politics or observes the different arrangements in family life of other branches of humanity. I know it is usual to overpraise the sea, to feel the despair of a long voyage when left alone with it, to curse the monotony of the view from the seaside lodging when we have ceased to curb our impatience of quiet; but still our thoughts travel back to our first love, and the rough health wafted from the ocean is not altogether replaced by the invigorating atmosphere of the hills. Beside which there is a stillness appertaining to the "everlasting" hills compared with the troubled waters of the ocean. Experience a night at sea with a night on the veld. The stars shine above both, there is the same silence, the same quiet; but there is a rigidity of thought amidst the solitude of the plains and hills compared with the poetic buoyancy produced by the sea. Amidst the solitude of the first our mind reverts to the genesis of creeds; on the water we breathe sonnets and listen to old Pagan music.

The summer of 1890-91 was remarkable for the heaviest rains that had occurred for many years. As we read in the papers of the phenomenal winter at home, so we were assured that the continuous summer downpour we experienced was equally unusual in South Africa. Towards the end of January the rivers were frequently flooded and dangerous, the roads in many places almost impassable, our homeward mails frequently missed the steamer at Cape Town, and our mails from home were uncertain of delivery. It was in this month that three Dutch anglers, who were sleeping in their wagons on the banks of the Pienaars River, within five yards of the stream, were swept away by a sudden flood or "coming down" of the stream, and the papers frequently recorded other fatalities from all sides. In February the bridge at the Six-mile Spruit was washed away, and all that month and during March fatalities to life and loss of property were of intermittent record. Outside the Republic the weather was equally bad; from Natal we heard that at Umbilo many Indian huts were destroyed and the Indians had to take refuge in trees, whilst in Durban itself we learned that at the end of one week in March for forty-eight hours there had been an exceptionally heavy fall of rain, the heaviest for twenty years. On the Sunday the people were practically weather-bound. There were no services in the churches in the morning, the streets and tram-lines were seriously damaged and the Berea tram-traffic was partly stopped. Perhaps the most vivid illustration of the effects of these river-floods in South Africa was obtained from Uitenhage, where one noon, whilst the Sunday river was rushing down with terrific force, the spectators on the bank observed in midstream a cart with two horses harnessed to it, dead, and dragging behind, as if fastened to the conveyance, was the body of a white man, which none could recognize as the ghastly flotsam sped swiftly to the sea. The last fall of rain before the dry season commenced occurred at Pretoria on May 12.

Flying all the year round is the ubiquitous butterfly *Danais chrysippus*, which is found over the whole of Africa, in South-eastern Europe, and generally distributed throughout Asia. *D. chrysippus* is also possessed of distasteful qualities which render it unpalatable to the usual insectivorous enemies, and thus affords an instance of a thoroughly "protected" butterfly. Its bright colour and slow flight show that it is subject to no fear of attack, or in the struggle for existence to which all living things have been and are, in a less degree, still engaged, this appearance and habit would have proved positive dangers to its long survival. It is not attacked by birds or other insectivorous animals, and is absolutely refused by them as food when kept in captivity. It is wonderfully tenacious of life, and specimens, after being pinched and pinned, have been seen, on the pins being withdrawn, to fly off in a nonchalant manner. Even when a dried and neglected museum-specimen, mites have avoided this butterfly, while they have destroyed other insects in the same box or cabinet-drawer. Its caterpillar feeds on a genus of Asclepiadaceæ (*Gomphocarpus*) which is everywhere abundant and also possesses distasteful qualities, so that its whole existence seems to be environed by natural *chevaux de frise*. How is it, then, that this insect does not positively swarm? is the question I frequently asked myself when watching the numbers which everywhere pursued this highly protected life. There must evidently be some great check at work, or the propagation of the species must result in prodigious flights, which would surpass anything to be seen in the whole Rhopalocerous order. I am inclined to think that these highly protected butterflies, which experience an immunity from attack on account of distasteful qualities or resemblance to some inanimate object or other protected insect, may have some inherent weakness or danger which produces great mortality in their early stages and that the wonderful protections we observe thus only enable them to escape extinction. This view would help to explain how it is that the extraordinary guises by which natural selection has enabled so many insects to escape the attacks of their enemies have not led to an enormous increase in their numbers. It is the weak that require protection, and like consumptive patients who live by escaping the rigors of a northern winter by visiting a warmer clime, but still possess the inherent weakness of their system, so nature grants these insects immunity from one danger, which allows them a possibility of surviving another. We know by the great gaps between the continuity of some species in the same genus how many must have reached oblivion in the struggle for existence, and I look on these "protected" insects as surviving by such means some incipient mortality of which we are at present ignorant, or their numbers must indefinitely increase. I carefully watched the *Danais* in the endeavour to find some danger to its life and the means by which its

Hemisaga prædatoria, n. sp.

increase was curtailed; but though generally unsuccessful, I did discover what I believe up to the present to be its only recorded enemy. This is a moderately large orthopterous insect (*Hemisaga prædatoria*, n. sp.), which I found lurking among the tops of tall flowering grasses, to which it has a considerable assimilative resemblance and which in this case enables it to secure its prey. The *Danais* hovers about, or partly settles on, the flowers and is then secured by the *Hemisaga*, which, in one instance, I found dismembering a freshly-caught specimen*. It is just possible, during the dry season, when insect-life is very scarce, that some insectivorous birds may, in a somewhat famished condition, make an experimental dash at a *Danais*. At that season I captured a specimen which was certainly mutilated as though by the bill of a bird, for the wings were not bitten symmetrically, as is the case when the attack takes place by a lizard or mantis, whilst the butterfly is reposing with its wings vertically closed †.

As is well known, the female of *Hypolimnas misippus* is a wonderful mimic of this butterfly. To an experienced eye the *Hypolimnas* may be distinguished from the *Danais* by its flight; but this is scarcely noticed without both species are known to be present and attention is thus directed. So close is the resemblance that well knowing both insects, I was not aware of the female *Hypolimnas* being present with the *Danaids* till I observed one *in copula* with its dark blue male. A purely English lepidopterist, not knowing these facts in mimicry could cross the veld and merely observe that *D. chrysippus* was very abundant. But these mimicking resemblances, by which the female *Hypolimnas* has found protection by being mistaken for the uneatable *Danais* and avoided accordingly, are even still more complicated. *D. chrysippus* has two varietal forms, *alcippus*, Cram., and *dorippus*, Klug, both of which occur in South Africa and both of which I found in the Transvaal.

* When in Natal that old lepidopterological veteran, Col. Bowker, informed me that he had frequently observed the *Mantidæ* to prey on butterflies.

† I am bound to affirm that this view, formerly advocated by my friend Prof. Meldola, was at the time contested by myself.

These varieties are very scarce, but both are also mimicked elsewhere by the female *Hypolimnas*. The same thing occurs in India, where, however, the mimicker of the var. *dorippus* is somewhat abundant, while the mimicked form is very seldom seen. Thus we have a butterfly mimicking a form which is almost extinct, and to a superficial observer weakening the theory which explains these anomalies. But it is necessary in all these cases to carry the mind back to the time when the butterflies, like all other living forms, were slowly establishing themselves by those qualities and appearances which, under the law of natural selection, enabled them to survive the struggle for existence. It was then that what we call "mimicry"—which is only one of a multitude of laws which govern the coloration of animals—first arose, and butterflies which slightly resembled uneatable species, or had somewhat the appearance of inanimate objects, would escape perils common to their kind, and these would thus become the dominant breed of the species, and be continually under the same selective process, till the disguise was almost perfect. If, then, we now find the present scarce form of the species so largely mimicked, it seems absolutely certain by the survival of the mimicry that it must have been once the dominant form of the species—at least in India—and has since, in the recurrent changes of nature, been almost replaced by the present form we so well know [*].

The food-plant of this butterfly (*Gomphocarpus*, sp.), which grows and blooms upon the most dry and barren parts of the veld as well as where moisture is found, is universally distributed in patches or small groups and is one of the earliest plants to spring up and bloom when the cold nights of the dry season become less severe. Its flowers are visited by many insects. From them I have collected some half-dozen species of *Cetonia* and some showy representatives of the Heteromera, as well as Gallerucidæ and Coccinellidæ. Many Diptera and Hymenoptera visit the bloom,

---

[*] For these views regarding the evolution of the species in India, I am largely indebted to Colonel Swinhoe, the well-known Indian lepidopterist.

and among Hemiptera a species of *Lygæus* is particularly abundant.

At about the end of November the shrill cry of the Cicadas was constantly heard from the willow and peach-trees in Pretoria, but principally from the first. The dominant species was *Platypleura divisa*, and I was surprised to find that it was captured and eaten by spiders. On once hearing a particularly loud chorus from a peach-tree, I visited the same to capture specimens, and found that spiders had industriously spread their webs between the branches, and remains of the *Platypleuræ* were suspended in a more or less devoured condition. I made use of these webs to procure specimens, for when first disturbed the flight of the *Platypleuræ* is wild and headlong, but by getting between them and the meshes of the spiders I was soon enabled to obtain what was required. It is reasonable to believe that these insects pair during their mature stages or breeding-season. We passed daily a small willow tree where I constantly noticed a solitary couple of the species, and this was also known by the fact that we drove them out on walking by and frequently endeavoured to capture them. The male was always tuning, and was probably addressing his mate. At length, unfortunately, the singer allowed us to capture him; the tree was henceforth mute, and I afterwards felt quite a remorse when my path took me by the then silent Cicadan home, for there was not the consolation of having captured either a new species or one new to my collection.

I was surprised to find how many living creatures one had known in Britain were also to be found in the Transvaal. In birds the European Bee-eater (*Merops apiaster*) and Montagu's Harrier (*Circus pygargus*) were not at all uncommon, whilst in insects one was continually meeting with some old friend. A List is appended to this volume of all these comrades one finds across the sea, or rather near the extremity of another continent; but with fixed ideas of geographical distribution, and our natural conception of an Ethiopian region, it was somewhat surprising to find that man alone was not the only migrant. My earliest English schoolboy days were recalled when I caught the Convolvulus Hawk-Moth (*Protoparce convolvuli*), bred the Death's-head Moth (*Acherontia atropos*), or gazed upon the numbers of the Painted Lady (*Pyrameis cardui*); whilst an earlier friend than all appeared with the summer rains. The Crimsoned-speckled Moth (*Deiopeia pulchella*) was a very old acquaintance. I had caught it in Surrey, met with it again in the Malay Peninsula, received it from Mogador, and now at the other end of Africa found it somewhat an abundant insect. The time of its appearance in the Transvaal is very protracted. I first captured it at the end of September, and found it still active on leaving the country in the following July. Flying in the strong sunlight, I have often mistaken it for a large *Lycænid*, as the pale azure-blue of the posterior wings is wonderfully reflected, and the red and black spots of the anterior wings are, during flight, scarcely, if at all, visible. Its flight is short and it is easily captured.

It was very soon after my arrival that I first saw the Secretary-bird (*Serpentarius secretarius*), that well-known African snake-eater of which we have all read. It is generally believed, and I was assured as a fact, that a £50 fine was inflicted for killing one of these birds; but in the Transvaal, as elsewhere, I soon found that the "*vox populi*" must be taken "*cum grano salis.*" I enquired of several well-informed men, including a newspaper editor, who stated that such was the fact; but at last I induced a friend on whom I could rely to make proper enquiries at headquarters, and after considerable trouble he discovered that there was no fine whatever on the statutes, but that a healthy and deterrent legend only existed [*]. Another legend appertaining to this bird and copied in popular books on ornithology is that its legs are so long and brittle that they will

---

[*] For this and much other reliable information I am indebted to my friend Meinheer J. H. E. Bal, of Pretoria, who has long been a resident in the country.

snap if suddenly started into a quick run. My man, Donovan, who accompanied me to the Transvaal, and, imbibing the zoological *furore*, assiduously spent his Sundays in shooting birds, procured me a very fine specimen of the species. By a long shot he broke its wing, when it made off at a terrific pace across the veld, followed by a spider and pair of horses as hard as they could go. Eventually it was come up with, and on the Kafir boy endeavouring to secure it, the bird showed fight, beat him off, and again started running across the uneven ground. My man now outspanned one of the horses and on its back galloped after the creature, which had obtained a long start. For more than three miles did this chase continue over the veld interspersed with ant-hills, and eventually it required the contents of two more barrels (buck-shot) to stop and secure it. This fact effectually disposes of its reported incapacity for violent running, as the hunt was over a long stretch of country of the most uneven surface. The crop of this bird was full of the remains of orthopterous insects [*].

But the bird of the open treeless veld is the Vulture (*Gyps kolbii*), and in places like the outskirts of Pretoria, where dead oxen and horses in some seasons plentifully strew the plain, these birds act the part of a sanitary board. A specimen I obtained weighed in the flesh 32 lbs., and as it was a full-grown example and a large bird, I think this may be accepted as the maximum weight. On days when none are apparently to be seen, if one carefully looks upwards towards the clear sky and scans the expanse, the diminished form of one of these huge griffons is sure to be made out, as from its lofty vantage it surveys a large tract of country. Should a bird be seen to alight, it is not long before others arrive from all sides and hover about the spot. There can be little doubt that high in the air these sentinels are always on

---

[*] Dr. Sclater informs me, however, that the legs of specimens confined in the Gardens of the Zoological Society are very brittle and liable to accident. The range of this bird is somewhat restricted. Emin Pasha did not meet with it on the East-African steppes, though he believed it existed there (' Emin Pasha in Central Africa,' p. 402).

the look out, so that the whole level country is thus under constant supervision, and when a bird is seen to descend or to be making off, that act serves as notice of probable quarry for miles around, like the early signal of the beacon-fire flashed from hill to hill. The usual sailing motion of the hovering bird is at once changed for a direct route, and its flight then, as far as I observed, was always four or five strong flaps of the huge wings, succeeded by a short straight motionless forward movement caused by the impetus thus obtained, to be followed by another four or five flaps as soon as the former motive power was exhausted. Usually shy, when gorged with food their habits are quite modified and they are easily approached. I once came across more than a hundred settled about two dead oxen. On each carcass were ten or twelve vultures at work, whilst the others in listless and gorged apathy rested around. The naturalist who has skinned a full-grown and *full-fed* vulture will not easily forget the operation. Now that the vast herds of game which once roamed over the veld are practically exterminated, the vulture becomes more dependent for its provender on the deceased domestic oxen bred by man, and the body of an ox is much preferred to that of a horse. Their food around Pretoria may become scarcer, as a movement was on foot to form a commercial company for gathering up these carcasses to boil down for soap. About the town gardens a bird almost as common as the sparrow in England is the Cape Wagtail (*Motacilla capensis*), but which by its tameness and partiality for the habitations of man reminded me of our robin, and, like that bird, is as little molested, save by boys, the natural enemies of all birds. Many entomologists have recorded the fact that they have never seen a butterfly attacked by a bird; but I not only obtained an Arctiid moth (*Binna madagascariensis*), which I surprised one of these birds in the act of killing, but also saw another actually pursuing a butterfly belonging to the genus *Acræa*, which is generally exempt from these attacks.

After an interval of some fifteen years Pretoria was

visited early in the month of May by a prodigious swarm of locusts (*Pachytylus migratoroides*)*. Travellers from the coast had passed through these devastating insect hordes, which apparently were working their way up from the Cape Colony. On the morning of May 11th our attention had been directed to myriads of locusts flying near the hills, and some few stragglers were

LOCUST-SWARM IN PRETORIA.

found in the town; but shortly after noon the air was darkened, as swarms only to be computed by billions came with a rushing sound over our heads and across

* The traveller Mohr met with similar swarms of probably the same locust on the banks of the Vaal River in 1869 ('To the Victoria Falls of the Zambesi,' p. 94).

our path. The light was obscured as with clouds of dust, whilst to walk through the flitting insects reminded one of the driving snow-flakes at home, as the pale hyaline wings and not the dark tegmina are observable during flight*. Stragglers continually fell out of the ranks, and we heard them drop on the iron roof of our dwelling. The flight was directed at different angles of one common direction, and stragglers constantly kept up a small counter-stream to the main body. The ground was thickly covered, and at sunset most of the flight had probably settled for the night. The heaviest portion of the main body, which might be described as the centre of the army, crossed us in about half an hour, but the flight continued long after and before. Their extraordinary numbers could be appreciated by the non-observable effect of their immense losses. Myriads were trodden under foot, our Kafir workmen collected them for food †, the poultry of Pretoria gorged themselves on their bodies. Two Crowned Guinea-fowls (*Numida coronata*) which I kept in confinement, and were always supplied with food, devoured so many of the locusts that I feared that they must die of repletion ‡; a large "Gom Paauw" (*Otis kori*) that we shot shortly afterwards had its crop crammed with the bodies of these invaders, but the great cloud seemed to suffer no diminution. On the next morning the ground was thickly strewn with the locusts; but they took wing as the sunlight became

---

* Carl Lumholtz was also reminded of a snow-storm whilst standing among a swarm of locusts in Queensland ('Among Cannibals,' p. 186).

† Holub, after eating these insects, felt he "could recommend a few locusts to any *gourmand* who, surfeited with other delicacies, requires a dish of peculiar piquancy; in flavour I should consider them not unlike a dried and strongly-salted Italian anchovy" ('Seven Years in South Africa,' vol. i. p. 199).

According to Livingstone, "locusts are often roasted and pounded into meal, when they will keep for months. Boiled they are disagreeable, but when roasted I much prefer them to shrimps, though I would avoid both if possible" ('Popular Account of Missionary Travels and Researches in S. Africa,' new ed. p. 31).

‡ Mohr's ostriches "ate locusts from morning till night, and four of them soon afterwards died of dyspepsia" ('To the Victoria Falls of the Zambesi,' p. 201).

stronger, and by the afternoon we were moderately free.

On May 25th we were again invaded, and again from the same direction. We had learned from travellers of the preceding day that another locust army was approaching, and a "transport rider" assured me that his oxen had refused to go on against the dense moving mass. This time the living cloud broke upon Pretoria about 10 A.M., and had virtually passed from us by 3 P.M.* This swarm was afterwards reported from Waterberg and Zoutpansberg, showing that its flight was in a northerly direction. In the early part of June, in crossing the Magaliesberg hills, I found them somewhat plentiful in a defile on the summit. This small colony were evidently stragglers from the higher portion of the flight and had thus ceased to form part of the main body, which was now some hundreds of miles in advance. News was brought down to Pietersburg from the Spelonken that the locusts had been so numerous as to prevent the informant driving a cart and four horses against them †. On the journey to the Cape in July I met with a considerable number near the boundary of the Republic, a larger swarm the following day about 50 miles beyond Kimberley, and another swarm about 40 miles further south. All these were flying northward, and would probably pursue the same routes as their precursors. This was my last experience of *Pachytylus migratoroides*. The year 1891 might be styled by entomologists a "locust year," for Southern Africa was not the only region invaded, and almost simultaneous reports were received from Egypt and India ‡.

As the colder and dry season commences the natura-

* Of this swarm a correspondent of the 'Transvaal Mining Argus' calculated that he passed through a cloud of locusts 25 miles long, about a mile and a half broad, and something under half a mile thick, giving about 12 cubic miles of locusts. Taking a low estimate he reckoned there would be about 2000 locusts to every cubic yard (an estimate much too low), and altogether he calculated that he must have passed over 130,842,144,000,000 locusts.
† 'Zoutpansberg Review.'
‡ 'Zoologist,' vol. xv. p. 221.

list can obtain many good specimens on the Pretoria market, for the Boers are then able to bring their game in for sale, which is impossible in the damp hot weather. The farmers are fond of shooting, but are equally glad to find a market for the game, which with forage, firewood, and other articles are sold by auction off the wagons before breakfast by the market auctioneer. Amongst birds, the Paauw (*Otis kori*) may often be bought, and I have known a heavy bird to fetch as much as £2 10s., for its flesh is very rich and highly flavoured, and I cannot agree with Mr. Ayres that the flesh " is too coarse and oily to be good eating "*. My man secured me a fine 20-lb. specimen, which he killed with No. 6 shot a few miles out from Pretoria. Its crop, as I have remarked before, was full of locusts, and it was certainly the fattest bird I ever skinned, my hands being saturated with grease by the time I had finished the operation. The bird does not seem at Pretoria to reach the great weight it does in other parts of South Africa. The proprietor of the hotel at which I boarded told me that the largest specimen he ever bought weighed 28 lbs., and a friend who had been an energetic sportsman for many years had only once bagged a Paauw that reached 32 lbs. On the other hand, I met a gentleman at Potchefstroom who said he had shot a specimen that weighed 41 lbs., and this was the largest he had ever seen or heard of in that neighbourhood. This I believe is about the maximum weight of which we have any authentic record, and I am somewhat sceptical as to the existence of the reported 50-lb. or 60-lb. Paauws †.

The smaller Bustards, *Otis carulescens* and *Otis afroides*, are not at all difficult to obtain on the market, and the Spur-winged Goose (*Plectropterus gambensis*) is

---

* Layard's ' Birds of South Africa,' Sharpe's edit. p. 633.

† Mohr states that he has shot specimens weighing thirty-five pounds (' To the Victoria Falls,' p. 33). Mr. Ayres, though he had often heard of 40-lb. Bustards being shot, never saw one of anything like the weight, though one of 40 lbs. was reported as shot by Mr. Buxton (Layard's ' Birds of Africa,' Sharpe's edit. pp. 632-3). Burchell describes his typical specimen as measuring in extent of wing not less than seven feet (' Travels,' i. p. 393).

rarely absent when game is brought in. Bucks of various species, the "jumping hare" (*Pedetes capensis*), the Monitor (*Varanus niloticus*), and skins of Leopards and the smaller cats I have also seen for sale. It is rare, however, to find a bird in good condition, as they are usually badly shot and with the plumage ruined. It is somewhat strange that the Boer farmers do not show more energy in bringing game to the Pretorian market, for it is certainly remunerative. During my stay a resident went on a shooting-expedition to the wood-bush about 90 miles from Pretoria, and on his return sold the game to a butcher for £27. Amongst the spoil were two bucks, two small paauws, ducks, partridges, and blue bustards, which at this price averaged 5s. per head all round. They were then retailed, blue bustards at 6s. each, partridges and ducks 6s. per brace, paauws from 30s. to £2. From this man I secured a very fine specimen of *Otis cærulescens*. All specimens, both living and dead, fetch fair prices in Pretoria, and a pair of healthy young Quaggas (*Equus quagga*) were brought in and sold during my stay for £55.

We occasionally obtained good sport among the so-called Partridges (Francolins), when the grass had died down in the early part of winter. The commonest species met with in the neighbourhood of Pretoria was *Francolinus levaillantii*, sometimes in good coveys, but never far away from water. These birds lie uncommonly close and can be easily passed. A Kafir boy once pointed out a grassy spot, not more than a yard or two square, where he assured us he had seen a bird settle down. We thoroughly, as we thought, threshed this spot, walking apparently over it again and again, and yet, subsequently, the boy with more perseverance and a desire to prove himself right, turned it up from under a tuft of tall dried grass that we had just missed treading down. Later in the afternoon of the same day, a single member of a covey which I had disturbed squatted in a small hole in the path about 80 yards in front of me, and depressing its back level with the earth, exhibited a good instance of the protection obtained by assimilative coloration. So perfect was the illusion, that partly owing to the diminishing light I failed to add it to my bag by a charge of shot.

The longer I observed living nature in South Africa the deeper became my impression that the colours and habits of the animals and plants around me were, like the geological contour of the country, a story of a bygone time. The colour of every feather, the appearance of every seed-capsule, is due to causes which in many cases are now almost inoperative. But it was then in the dire struggle for existence, subsequent to the last great geological change in the surface condition of the earth, that those varieties of plants and animals only survived which could in some way pass the severity of a competitive examination by natural selection. Hence we must not always expect to find a philosophical explanation of the bizarre colours of animals and plants by simply considering their present conditions of life. If it is difficult to trace the evolution of a civilized community of mankind, with its customs and superstitions, to its primordial elements, many of which belong to a prehistoric period, how gigantic is the task to attempt to go behind the very evolution of man himself! and yet it was at that time when the small birds and insignificant insects obtained the maximum of their colour-markings, not to add to the beauty of the scene, but to enable them to survive an eliminating process which took place in the great struggle for existence. Many of these gorgeous living forms are to my mind fossils, of a past epoch which we cannot read.

THE MONITOR (*Varanus niloticus*).

CHAPTER V.

THROUGH WATERBERG.

Scarcity of timber in the Transvaal.—Leave Pretoria for Waterberg.—Waterless region of the Flats.—The Warm Baths.—Beautiful scenery.—Euphorbias and their poisonous qualities.—Fever districts.—The Massacre at Makapan's Poort.—Sanguinary retribution at Makapan's Cave. — A fine orthopterous insect. — The Prospector. — Reptiles.—Ravages of the "Australian Bug." — Majuba day. — Mimicking insects.

EARLY in the month of February I made a journey through the Waterberg district, to procure a supply and estimate the quantity that could be obtained of the best tanning-material of the country, the leaf of the tree I have already referred to (*Colpoon compressum*). As the industry of the Transvaal progresses, an investigation of its tanning-products will doubtless be undertaken, for it can scarcely be credited that the few vegetable materials now only known as available for a trade that must have a future adequately represent its wealth in this matter. A process of tanning in small quantity for household wants has long been understood and practised by the Boer farmers, who, I am informed, use the leaves of other trees than the *C. compressum* for the purpose; and when the government really takes up the necessary question of forestry, a most important one for the country, the preservation and planting of these trees, upon which will depend the success of a Transvaal leather manufacturing trade, must be seriously dealt with *.

At present the Transvaal may almost be described as timberless, so far as building-operations can be carried on. Even the wagon-builders have no local material, or at least none that can be obtained in any quantity, and it is absolutely cheaper to import wagons from the British Colonies, where there is an official Inspector of Forestry, than to manufacture them in the heart of the Transvaal. Vast quantities of deals and other European and American woods are brought up from Durban with all the incidental cost of rail and ox-wagon †; and when at last the railway is allowed to give to the development of the country its natural and much-desired impetus, the sleepers for the lines will have to be imported. At present the great drawback to all local industries is that articles, despite duties, and in the face of monopolizing concessions, can be imported as cheap or cheaper than they can be manufactured on the spot. The wealth of the Transvaal has hitherto only been sought beneath the ground; it must now be cultivated on its surface.

I started just after a period of heavy rains, and as the coach passed through the Wonderboom Poort, signs of the recent floods could be observed by the vegetable

---

* This has been thoroughly done in Australia, and Mr. Maiden, in his 'Useful Native Plants of Australia,' has described over thirty species of "Wattles" and about half as many Eucalypts which have been tested for tanning-material. In all eighty-seven Australian species have been under examination.

Burchell found that the Hottentots used the bark of the Karro-thorn for tanning sheepskins, and amongst other plants used for the same purpose were a kind of *Ficus* and *Mesembryanthemum coriarium*, B. ('Travels in Interior of Southern Africa,' vol. i. p. 243).

† In October 1890 the following quotations were obtained:—Deals, 3×9, 1s. 5d. per foot; flooring, $\frac{7}{8} \times 6$, 4d. per foot.

débris left stranded in the tops of trees growing by the banks of the river; many of the trees were at least fifteen feet high, and one could thus realize the dangerous and relentless force of these flooded streams. After leaving this mountain pass, short scrubby trees become plentiful, and the soil is loose and sandy. As the journey is advanced, the country is found to be much more wooded and is in pleasant and strong contrast to the monotony of the bare veld which marks the higher lands. To drive along a narrow road through thick woods was, indeed, a novel experience, and we reached the banks of the Pienaars River about 4 P.M., and shortly afterwards commenced the longest and most severe stage of our journey. The "Waterberg Flats" occupy a waterless region of some twenty-five miles in width, where there are no stages, and the mules have at least a four hours' stretch; but on this occasion, owing to the state of the road, in which the wet rutty ground had dried just sufficient to be bad for the feet of the mules, we were five hours in transit. During the last hour one could not help sympathizing with the poor jaded beasts, and the shouts of the driver and the crack of the whip were constant sounds. Passengers and mules will probably soon be spared this unbroken stage, as an enterprising American was then sinking a well, already 108 feet deep, through the rocky ground at his own expense. When he reached water, and had completed the well, he proposed building a store and stables, and as the spot is about midway across the Flats, his enterprise should be repaid. It was 9 P.M. when we reached the hotel which bears the name of the Warm Baths. The warm water rises from a mass of peat and reeds in the neighbourhood, and is conveyed to the hotel by pipes. After the dust and fatigue of the road these baths are most refreshing, and now that the property is leased and managed by a small British company in Pretoria, the spot bids fair to be the retreat and sanitarium of the capital. The Boers visit this spot and use the waters; but in their case a hole is made in the ground, into which the water flows. In this rude and muddy bath, covered with a tent or other screen, the farmer will remain for hours.

A real night's rest is quite an unknown quantity to the coach-passenger; this journey was, of course, no exception to the rule, and we were aroused at 2.30 A.M. to resume our route. By 7 A.M. we had reached Nylstroom, a forlorn spot, where the imposing appearance of a post-office and landdrost's court, unsurrounded by any apparent business life, give it the appearance of a still-born township. But fever has been the retarding cause of Nylstroom's future, and its character for unhealthiness will long survive, though the natural beauty of the surrounding country, and its little-disturbed condition, should make it a district beloved of sportsmen. As the traveller leaves this spot it is difficult to believe that one is still in the Transvaal, after an experience only of the country between Pretoria and the Cape and Natal frontiers. Woods, park-like tracts, undulating country, from which views could be obtained of endless and varied landscape, tall, wooded, isolated hills, and ranges of mountains with forest slopes, alternately meet the eye. Scattered Euphorbias quite transformed the appearance of the flora, and broke, as it were, the sameness of the short forest growth. The irritant properties of the milky juice obtained from these plants is well known *; but the bloom possesses the same attributes, and honey is unfit for use that has been made by bees which have visited the flowers. A resident friend once purchased some honey from Kafirs, and this, when used by himself and companions, caused an intense burning sensation in the throat; they then made careful enquiries as to its origin, and traced it to a Euphorbian source. New birds were observed in the trees such as never appeared at Pretoria. A hornbill was common, but more abundant still was the pretty Lilac-breasted Roller (*Coracias caudata*). At intervals on the tops of trees perched Buzzards, that seemed by their numbers to have the whole neighbourhood under

---

* Used by some of the tribes of South Africa for poisoning water to obtain game (Parker Gillmore, 'Days and Nights by the Desert,' p. 61).

observation, and yet when I traversed the country again about two months subsequently, scarcely one of these birds was to be seen. A large portion of the avifauna is migratory in a local sense, and the Buzzards follow their prey.

We now approached localities which will always be remembered in Boer history and recall the days of the Boer and Kafir struggle for supremacy. Potgieter's Rust is associated with a name attached to a tragedy about to be related. The place had been an improving hamlet, and had enjoyed a healthy reputation till the year 1870, when fever in a most violent form broke out among its inhabitants. By April of that year eighty-one out of the ninety-three settlers had died or were prostrated, and in May the locality was deserted. It is now again inhabited, and may in time become a township. A spot, however, which is still called Makapan's Poort, is the central point of one of those wild deeds which so often give a lurid glare to the struggle between native races and white settlers. At Makapan's Poort, in the year 1854, a particularly diabolical murder was perpetrated by a clan of Kafirs under a chief named Makapan upon a party of hunting Boers. The hunting party consisted of thirteen men and ten women and children, and were under the head of a Field-cornet, Hermanus Potgieter. Potgieter had visited Makapan to trade for ivory, although the volksraad had passed laws prohibiting this manner of barter, with the view of preventing the danger of disputes and quarrels arising between the black and white people. Whatever the provocation may have been in the demeanour of the Boer, if provocation there was, as has been currently reported at the time and since, it remains that these unfortunate people were barbarously murdered, women and children sharing the same fate, and Potgieter himself flayed alive, his skin being afterwards prepared for a kaross.

Blood once being shed and the die cast, the Kafirs commenced to pillage the surrounding neighbourhood. Needless to say the fiercest passions for retribution were now aroused among the Boers, and a sense of danger demanded a swift reprisal; no homestead was safe if this Kafir attack was allowed to develop, every farmer instinctively apprehended the emergency, and soon upwards of four hundred armed burghers had arrived at the scene of the tragedy determined on vengeance deep and terrible. The Kafirs fled to a huge cavern some two thousand feet in length and four or five hundred in width, which was closely blockaded by the Boers. Now commenced that wild revenge which is common to man's nature under similar circumstances; it has been practised by the French in Algeria, and by ourselves during the Sepoy revolt in India. Frantic with thirst the imprisoned Kafirs sought at night to reach the water that flows near the cave, but were shot down in the attempt; quarter was a word unknown, and after twenty-five days' blockade, the cavern was entered and its horrors seen. According to Commandant Pretorius—who would have no interest in exaggerating the figures—nine hundred Kafirs had been killed outside the cavern, and more than double that number had died of thirst within it [*]. Makapan himself is reported to have perished by poison introduced in water, but the true story of the wild vengeance will probably never be told. It was during the blockade that the present President Krüger exhibited an act of that bravery which he has elsewhere displayed. A Boer commander was shot when standing near the mouth of the cavern, and Mr. Krüger volunteered to bring away the body, which he did. This man was afterwards buried on his farm, and I have visited the grave; it was silent and alone, as befitted the last rest of an old voortrekker.

Some eight hours were at my disposal before the return coach could convey me back to Pretoria, and I seized the opportunity to visit the cavern, guided by one who knew the neighbourhood and had once been

---

[*] The South-African historian, G. McCall Theal, who is cautious and not biassed against the Boer, adopts these figures ('History of South Africa, 1854-72,' p. 30).

an English soldier. The weather was clear and hot, we crossed large fields of maize grown by Kafirs, who are here the only agriculturists, and as we walked

Clonia wahlbergi.

through these high and flourishing plants one was reminded of the fields of young sugar-cane in the East. It was in these fields that I first captured the fine orthopterous insect, *Clonia wahlbergi*, and experienced the severity of its bite. I had previously sustained the pincer-like grip of the beetle *Manticora tuberculata*, which was much less painful than that of this Orthopteron, the mark of which on my finger was carried for several days. An hour's walk brought us to the first cave, which the Kafirs visited before proceeding to the second and larger one, where they sustained the blockade and in which most of them perished. It was very hot, and when we reached the abrupt rocky side of the hill up which we had to climb, for the cavern is situate some distance from the base [*], we were glad to quench our thirst at the small stream of cold clear water that flows along the valley at its foot. It was this stream that the thirst-maddened Kafirs sought to reach at night, when, however, the Boer bullet was usually received. Inside the gloomy precincts of the cavern skulls were strewn in profusion, but generally without the lower jaws, and many have been taken away by visitors: the dung of the sheep and goats possessed by the imprisoned Kafirs was still intact on the dry floor, and handles of axes, grindingstones for corn, baskets, &c., bore their witness to the retributive slaughter of 1854. We could not penetrate into the recesses of the cavern, as we had not brought candles; but it was an uncanny scene, and a large dog that accompanied us seemed very ill at ease and kept near the entrance. I was able to select six very fair crania [†], both juvenile and adult, which I brought away, and we retraced our steps, glad to reach the "hotel" once more and drink a bottle of English ale, which, however, in this part is priced four shillings and sixpence [‡].

All the way, both coming and going, we saw the

[*] Mr. Alford describes these caves, of which there are a number in the neighbourhood, as "large water-worn cavities in the stratification of the quartzites, formed by the removal of portions of the softer beds" ('Geological Features of the Transvaal,' p. 49).

[†] These crania are now incorporated in the fine craniological collection belonging to the Museum of the Royal College of Surgeons; and are fully described by Professor Stewart in the Appendix to this book (see p. 157).

[‡] In the Spelonken I once paid 5s. 6d., which may be taken as the highwater price for our English beverage.

commencement of the Mashonaland trek. Wagons, drawn principally by donkeys, well equipped, were bearing young and enterprising spirits to Rhodes' new country and England's new Protectorate. Prospectors were hastening to find and peg-out claims which contained the precious reef, and though much fever and more hardship will be encountered in the early days, it will probably be the South-African land for the future colonist and will remain under the old flag. It is bound to absorb some of the capital of investors which might have otherwise reached the Transvaal, and though Boer and Hollander may sometimes think the Republic can do without the English, it will still miss the influx of English money.

When a man has once gone prospecting he finds a charm in the life which he seldom deserts. Of course I am speaking of those free spirits who are no use in business, have a moral law unto themselves, and love the solitude of nature, diversified by an occasional carouse in a large town. Such a one we carried in our coach on the up-journey. He was bound for Mashonaland, and had purchased the wagon and oxen to carry the party, his friends having contributed the other necessaries. The wagon, however, had gone on without him, as he informed us he had indulged in such a "paralatic drunk" that his friends had become tired of waiting, and he was now endeavouring to overtake his party. Another member of the staff had still to follow. Four times had this susceptible man driven to the Poort where the wagon waited to start, and each time accompanied by a "lady" friend to see him off, but on each occasion his will failed and he returned to town with his fair companion. These men when they do get out and settle down will be sober slaves, but they are like sailors on shore when a town is reached. My companion was a lump of good-nature, of strong build and constitution, and in all my experience at home and abroad I never saw a man drink so much and show the effects so little. Consequently what the nature of the banquet was which prevented his joining the wagon, can be more easily imagined than described.

The rain again commenced on our return, and we found Pretoria once more a scene of mud, with the usual results of detained mails and an almost impossibility of heavy transport. The arrival of the weekly mail is to the European exile an event of the first importance. Seven days' intellectual stagnation, in which the only recorded events are found in the dismal swamp of Boer politics, Church squabbles, and mining reports, render the new home journals most attractive, though the contrast—apart from purely literary studies—is the record of the same motives being applied to a larger and more complex field. The same elements that compose the social and political fabric of Boerland are found at the root of our own national institutions; but in Europe the stage is larger, the principal parts are acted with more dignity, and the scenery and decorations more impressive. The subject matter is the same, but the oratory has been more developed at home; the "Oom Paul" of the Boer and the "People's William" amongst ourselves represent only a difference in degree and not of type. So it is with the civilization of Pretoria, which has reached a gaol and permanent gallows, but not yet acquired a workhouse; it has recognized crime, as ours has, but still lacks the accompanying abject poverty of our own more developed towns and cities. Thus our home papers recorded the acting of a great national tragedy or superb social comedy, whilst the Transvaal existence has only yet advanced in politics to an ordinary drama, and in social distinctions to a farce. Midas may arise and does appear in the Boer republic, but he has not the potentiality for display which the plutocrat possesses in Europe, and appears ridiculous where our own creations are sometimes only offensive; it is the difference between the processionary splendours of a travelling circus and the more gorgeous vulgarity of a Lord Mayor's Show. Thus a long and late night with the London papers was always a weekly treat compared with the uninteresting records of Transvaal communities; but how different the impression became when leaving the townships one once more visited the solitudes of

South-African nature, and then the petty aims and sordid cares of our boasted development appear like an agony or a nightmare. The young Briton without family ties at home who has once roamed over these wild plains, and lived the free life, will visit, but probably never die, in the old country. The anomalies of our so-called civilization are seldom really experienced or so clearly seen as from the vantage-ground of Nature's solitudes, and we there learn a lesson which we never forget, and acquire habits which last for a lifetime.

Reptiles are not abundant in the neighbourhood of Pretoria. Lizards (*Mabuia trivittata*), which live in holes on the banks and hillocks of the veld, may be often seen in fine weather curled up at the entrance to their abodes apparently enjoying the air; they then arrange their bodies in a circular manner, their legs falling flat by their sides, and thus have all the appearance of snakes. They were difficult to capture without shooting, and I frequently dug them out, when they were always found living in the company of Toads. A Monitor (*Varanus niloticus*) was not uncommon about the banks of the spruits which here and there intersect the veld; in the stomach of one I found the remains of two freshly-devoured rats and a frog. Among the different carcasses brought to the morning market by Boers for sale maybe frequently found the body of one of these animals. One Lizard (*Agama hispida*) is not at all uncommon, and I have secured three or four specimens from under one stone. Snakes are certainly few in number, and no Irishman need fear meeting too many specimens of his pet abhorrence near Pretoria. The Python is scarce; I heard reports of solitary examples having been seen in widely separated spots, but was unable to obtain a specimen. One of the most mythical animals is the Crocodile, which is often reported as inhabiting streams which certainly do not possess a single individual. This was particularly the case in the Spelonken, where I was prevented from bathing in the deepest and best pools of the river by reports of these Saurians, of whom none of my informants had ever seen a specimen at the spot.

As the untutored mind is apt to people the air with ghosts and goblins, so the Kafir loves to imagine the waters of the dark stream as inhabited by river gods and great reptiles. Even sailors find it difficult to believe that the vast silent ocean is not peopled by huge sea-serpents and other monsters; but, alas! all things, and even fancies, die a natural death, and the sea-serpent has now nearly followed the mermaid. Zoological science has made it impossible to

> "Have sight of Proteus rising from the sea;
> Or hear old Triton blow his wreathëd horn."

The dangers of these rivers are not from their inhabitants, but from their swollen and sudden rush during the heavy rains. We once narrowly escaped in driving through one of these augmented streams. The water rose over the floor of the spider, which floated, and for a few moments the horses lost their foothold; but I shouted to the Kafir boy to use the whip, and we got through. The Boer farmer I visited would scarcely believe we had driven through the stream (which was certainly due to ignorance and not courage on my part), and on our return he sent two of his sons to the river to help in an emergency, or to witness a foolhardy Britisher have at least a dangerous ducking. Of course under such a challenge the thing had to be again attempted, and we succeeded in accomplishing our purpose, though with an unexpressed resolution to try no such experiments again. The Kafir boy showed no fear, nor did he on another occasion, when the horses breaking from control took fright in going down a rocky hill and bolted, while for several moments I was asking myself whether it was to be broken limbs or broken neck!

Although, as before remarked, the high veld is an almost treeless region, and Pretoria by planting has been made an exception to the somewhat general rule, its arboriculture is in danger by the arrival of the Coccid, or so-called "Australian Bug" (*Icerya purchasi*), which has ruined many trees and shrubs. Already a formidable pest in Australia, New Zealand, and North

America, it was first observed in the Botanic Gardens at Cape Town in 1873, and has since spread over nearly all South Africa, this scale-insect being now too frequently seen in the Transvaal. It specially attacks the orange-tree, which in the high Transvaal is the only really eatable fruit to be obtained, and hence its arrival and depredations are the more to be regretted. This Coccid* in time may prove as serious a trouble to the arboriculturist as the prevalent lung-disease already is to the cattle-farmer and the horse-keeper. Man's development of this country is a long struggle with the different forces and agents of Nature; if his cattle survive the sickness in the Transvaal they will not conquer the little Tsetse-fly (*Glossina morsitans*) of the interior; heavy rains and floods destroy his crops, and the scale-insect attacks his trees; in the rich lowlands, where the most luxuriant crops can be produced, malarial fever dwells; in the townships of the healthy highlands defective drainage is attended by malignant typhoid epidemic. Man's greatest happiness is living in conformity with Nature's laws, his greatest intellectual achievement has been in conquering and utilizing her forces. Dynamite is a progressive power in the Transvaal, and is an invincible force in hewing the railway-track through the quartzite rocks, constructing roads across adamantine defiles, or blasting the gold-bearing reefs. The boom of its explosion is a sound often heard, always denoting industrial enterprise; and the word dynamite had a strange significance in my ears in this land as I observed its destructive force utilized for constructive purposes, and remembered its felonious notoriety in London a few years previously.

---

* To those who would consult the literature relating to this insect, its life-history, and the proposed preventive measures against its attacks, may be recommended the following works:—' Report of Prof. C. V. Riley, Entomologist of U.S.A., for 1886,' Washington, pp. 466–492;—' Insects Noxious to Agriculture in New Zealand. The Scale Insects (*Coccididæ*),' by W. M. Marshall, 1887;—Report by Mr. Roland Trimen—Government Notice (Blue Book) No. 113, 1877;—and lastly, the excellent *résumé* on the subject by Miss Eleanor A. Ormerod in her 'Injurious Farm and Fruit Insects of South Africa.'

February 27th, or Majuba day, is rightly remembered by Boers as a general holiday. Englishmen can accept a defeat, but need not necessarily celebrate its anniversary, and with my nephew and man, who had accompanied me from England, I started on the previous evening for the small quantity of "wood-bush" that may be found in the Pretoria district on the Waterberg Road. An old colonist, who had reached Natal as a child, and wandered about South Africa ever since, often deserted but never quite forsaken by fortune, who seemed to have never failed and never prospered, and who, without any great financial reputation, was content in disposition and seemed independent in character, invited us to spend the night at a small farm he rented in the neighbourhood. We reached the abode late, for the way was long, the roads heavy, and the night dark, and here in this small domicile on the vast veld, dwelling in all the plainness of the most primitive farm at home, was a colonist family who only just preserved in the parents' early life the slightest touch with home. And yet it is with these good people that the distrust of the Boer is most strongly felt. The wealthy colonial or British merchant thrives with the Boer and respects his customer, but with men of small means and plain living the difference is most pronounced. The soldier accepts and forgets his defeat, but these humble and industrious Scotch and English, who were scattered over the country with farm or store at the time of the war, and went through much danger, and, what was worse, had to put up with much rudeness, have, doubtless, forgiven but certainly not forgotten. Our sleeping accommodation was at least primitive: a straw paillasse stretched on the earthern floor of an empty cowshed, which, nicely ventilated by holes in the walls and roof, and agreeably perfumed by strings of onions suspended from the rafters, afforded us, in the absence of rain, excellent shelter. I was informed here of a sudden cessation to a bird pest. A small Finch had swarmed on the farm to the great destruction of certain crops, and all attempts to destroy them or thin their numbers had failed. These birds roosted at night on the reeds growing in a small river-bed or vley, and one night, shortly before my arrival, and after particularly heavy rain, the waters suddenly rose and covered the reeds to the almost total destruction of the birds, for my host said he had seen scarcely any since. We frequently see swarms of insects swept away by floods, but I had not hitherto heard of a wholesale destruction of birds by the same means.

The wood-bush we visited was only a few miles in extent and thin in appearance, and yet contained almost another zoological world to the bare veld which adjoined it. Birds of many species not seen before were now met with, and many new skins secured for the collection. In insects the fine day-flying Moth (*Xanthospilopteryx superba*) flew amidst the shade of the acacias, and in Butterflies *Herpœnia eriphia* and *Teracolus eris* and *T. evenina* were captured by myself for the first time. The fine Ant-lion (*Palpares caffer*) was abundant round the outskirts of the trees, and large and gaudily-marked Spiders (*Nephila transvaalica*)* occupied in family groups or industrial communities the immense webs that stretched from tree to tree. In the ardour and pleasure of collecting we had aimlessly wandered among the trees, with the inevitable result that about noon we found we had not only lost ourselves, but all held different ideas as to the direction we should pursue. It is at such times that the mind grasps the full benefit of both savagedom and civilization, for we possessed neither the wood-lore nor path-finding capacity of the first, nor did we carry the pocket compass of the latter. Of course we went miles out of our way, and after hard walking for hours under a broiling sun we at last reached our spider again, and arrived late in Pretoria on the evening of Majuba day.

Since January our Coleopterous visitants had included the fine and showy Buprestid *Sternocera orissa*. The first time I saw this grand beetle—for in the Transvaal the Beetles, as a rule, are neither large nor

* A new species, described in the Appendix by Mr. Pocock (Tab. V. fig. 4).

showy—it was resting in some numbers on the still leafless branches of a solitary acacia on the bare veld. Being far beyond our reach we threw large pieces of quartzite against the branches, and the concussion, as a rule, brought the insects to the ground, when they were secured before they could take wing. This species was always found on the branches of an acacia. Beetles are, however, difficult to obtain; they are plentiful for a short time at the commencement of the rains, then become scarcer as the summer season advances, and are almost totally absent during the long dry season. Although the hedges were a mass of roses constantly in bloom during the summer, I was surprised to see how little they were visited by floral beetles. Certainly myriads of the Cetoniid *Pachnoda flaviventris* could generally be seen, and also the large Cantharid *Mylabris ophthalmica*, but the majority of all these flower-visiting Coleoptera confined themselves to the small and obscure bloom found on the veld. A new tree would burst into bloom, its flowers lasting but a short time, during which frequently a species of the Cetoniidæ not hitherto seen would visit in quantity this fugitive blossom and again quickly disappear with it.

From long observation in the field and of the contents in my cabinets at home, I had become convinced of the phenomena and the truth of the theory of mimicry[*] in the insect world, by which under the law of natural selection edible species showing any resemblance to inedible ones, have gradually been preserved by the protection thus afforded, and the same selective process going on among their progeny for long periods of time has resulted in those wonderful resemblances which we now find among distinct orders of insects. So strongly was this always in my mind that I frequently was stung by real Hymenoptera, when I expected too much and thought I might be handling an imitator. But the tables were quite turned when I first captured a female of the longicorn *Amphidesmus analis*, which on a leaf has a surprising resemblance to a female of the genus

[*] Long since enunciated and proved by my friend Mr. H. W. Bates.

Lycus belonging to a totally different Coleopterous family, and I was completely deceived till I held the insect in my hand. The objections urged against the theory of mimicry are generally based on a total misunderstanding of the theory itself. One frequently listens to arguments against a hypothetical assumption that an insect of its own volition, for protective purposes, copies the garb and appearance of an inedible species. Such a wild proposition would require no objection, for it could obtain no support. It is only when one has realized the struggle for existence in all animal life—including man himself,—has recognized the unbending, inexorable, and universal application of natural laws, appreciated that benevolence is an acquired product of the human heart and not of natural life, and observed that all life exists in an iron-bound environment, where strength reigns supreme and the strong taketh by force—it is only then one understands what Herbert Spencer has so well called the "survival of the fittest," and what Darwin had enabled him thus to see by his enunciation of "Natural Selection." With these facts before us we can comprehend how this "breed" of the persecuted beetle, ever tending by the attacks of its enemies—a form of natural selection—to perpetuate its race by its more favoured representatives who were mistaken for inedible species, in the course of time reached—in scanty numbers, it may be—its zenith in simulative appearance and escaped extinction. These mimicking species are the shadow of a past, when there was a great need and a great danger.

NATIVE HUT, SPELONKEN.

CHAPTER VI.

ZOUTPANSBERG AND THE MAGWAMBAS.

Start for the Spelonken in Zoutpansberg.—Horse-sickness.—Pietersburg.
—A fine Convolvulus.—A castellated residence in the Wilds.—Night
in a wagon.—Kafir traders.—Kafirs on the tramp.—Polygamy.—The
Magwambas, their customs and institutions.—An ox feast and dance.
—The Makatese.—The Mavendas and their iron-work.—Birds' food
largely orthopterous.—Good entomological spots.—Zoutspansberg with
its natural riches still undeveloped.

I HAD for some time intended to undertake a journey through the Zoutpansberg district, and was engaged in making enquiries as to the best mode of conveyance to be engaged at the termination of the coach service at Pietersburg, when I was introduced to Mr. G. D. Gill, a Spelonken trader, who kindly invited me to share his wagon on his return journey, and to accept his hospitality during my stay in his neighbourhood. We started for Pietersburg on a cold April morning, and although the coach was timed to leave at 5 A.M., it was not before another hour had elapsed that our black driver appeared upon the scene, when to the repeated and somewhat energetic remonstrances of the coach proprietor he merely returned the soft answer: "No, Baas, it cannot be six o'clock, I am sure."

The horse-sickness was now prevalent; a few days previously, when travelling to Johannesburg, we had to unharness a horse and leave it on the veld; on this occasion we soon had to dispose of one of our mules in the same manner. The number of animals lost by the coach proprietors owing to this epidemic was something enormous. Within the few weeks previous to my journey, the small regiment of State artillery had lost twenty-five "salted" horses, and the detachment of ten men which escorted the President to Natal were deprived of four animals between Pretoria and the Transvaal border. At present there is little or no cure known for this disease, which is a serious matter for the welfare of the Republic.

The journey through Waterberg has already been described in the previous chapter, and soon after leaving Eyting's "hotel" the country once more resumes its treeless and uninteresting appearance. We reached Pietersburg about 10 P.M., on the second day after leaving Pretoria, calling at Smitsdorp and Marabastad on our way. Pietersburg is a township now in course of development; it is planned out with sites reserved for Market and Church Squares, as in Pretoria and the other Transvaal towns. Already three churches were either quite or nearly completed; it also possesses a Landdrost, has an exceedingly healthy and open situation, is the market town of Zoutpansberg, and as Mashonaland prospers, Pietersburg must grow, for it is the last Mart on one of the principal roads to Rhodes' eldorado. Its principal inhabitants are Germans, its stores trust to a Boer trade; and though the first prosperity of Pietersburg had received a check at the time of my visit, the township has a future. Erven, or plots, that could have been purchased a few years since for £14, are now worth from £200 to £300. We stayed a day here waiting for our wagon, and time passed very slowly; there is nothing to be seen or done in Pietersburg but business, and at the time of my visit very little of that was acknowledged. The scenery around is bare plain and mountain, and health may here be restored at the cost of much *ennui*. It was difficult to realize that this was once a great game country, and living Boers can still remember the time when not only bucks and antelopes abounded on the now silent and lifeless veld, but even giraffes, lions, and elephants were to be found. Animal life was now almost alone represented by large numbers of the White-bellied Crow (*Corvus scapulatus*), which were more numerous here than in any other part of the Transvaal I visited, and the scanty flora was made memorable by a cultivated Convolvulus with blooms twice the size of the ordinary *Convolvulus major*, which was also most abundant in gardens. I saw this fine flower again in the Spelonken, and obtained seed from it, but I have as yet been unable to effect its germination in England.

After a day passed in Pietersburg, we started in a small wagon drawn by eight oxen for the Spelonken area of the Zoutpansberg district. The first day's trek was over bare veld, and towards evening we passed one of the most incongruous sights I saw in South Africa. Here in the desert plain suddenly appeared an effigy of an old feudal castle, reminding one more of a stage effect than of an antiquated building. This extraordinary structure has been built by a retired native commissioner, Capt. Dahl, and here he proposes to dwell and, I believe, end his days. I never fully realized before the true horrors of false taste; here where a bungalow with flowered trellis and garden rich in native flora would have harmonized with the natural features of the scene, we found a second-rate representation of what was most hateful in architecture and inconsistent with its surroundings. We rested at sunset near the base of a range of hills and then trekked on till about 11 P.M., when we again outspanned the oxen and passed the night on the outskirts of a field of Kafir maize. The first night passed in a wagon has all the charm of novelty; as one gazed through the opening behind at the clear starry sky, and realized the quiet of solitude, it seemed as though life was at last free, and social existence deprived of its fetters. With the second day's trek

CASTELLATED RESIDENCE IN ZOUTPANSBERG.

the scenery altogether changed, the country was more or less thickly wooded, especially after fording the Dwaas River, which we reached about noon. A few hours from this spot we crossed a plain studded with granitic hillocks, which rose like rocks and islands from a shallow sea; viewed from above, the whole scene reminded one of some portions of the coast of Brittany at low water, and it was difficult to overcome the impression that we were gazing on an old ocean-bed. Most of these elevated masses of granite were quite bare, with their surfaces highly heated by the rays of the sun.

The only Europeans we met on our road up the Spelonken were the traders, who keep Kafir stores. They all seem to succeed, and some are moderately independent after years of patience, industry, and solitude, for their life is a lonely one, especially when they are,

H

as in many cases, unmarried. The living is bare and but little diversified, home comforts are in some instances of the fewest number, whilst in the small flower-garden near the house may frequently be seen the tomb of some loved one, who has lived and died with them in these African solitudes, and whose remains now really consecrate the ground. Then, again, there is much leisure time, for the Kafirs come to purchase in a sporadic manner, and hours pass without the visit of a customer; consequently these hermit merchants are glad to have a chat with any passer by, and I found them very hospitable to me on my journey. An old Matabele trader named Cooksley, whom Mohr mentions in his Travels, has now settled here and has the best establishment on the road. The beauty of the spot is its flower-garden and orchard, both of which are due to the horticultural taste and industry of Mrs. Cooksley, who kindly supplied me with a stock of fine oranges on both upward and return journey; it is such cultivated spots and well-kept homes that are required to be distributed among the districts inhabited by the Boer farmers, for nothing but a healthy emulation can arouse that lethargic stock. These traders altogether depend upon their native customers, and in return are able to afford them considerable protection, particularly if they happen to be in the hands of unscrupulous and oppressive native commissioners. I heard many reports of savage floggings and impositions when I was in Zoutpansberg; and the government should remember that officials do not become valuable only as they collect native taxes, for it is possible at the same time to drive Kafirs from their locations, and thus not only destroy a source of revenue, but also depress a very valuable branch of trade in the country *.

All along the road we passed small bodies of Kafirs tramping home after working at Kimberley, Johannesburg, or Pretoria, where they usually remain for a

* The government quite recently instituted an enquiry into these charges, which could not be substantiated. It was admitted, however, that the Commissioner had flogged a native servant girl with a riding-whip for "frequent acts of immorality," but her subsequent death was decided to be due to other causes.

period not longer than three to five months. By that time they have saved a few pounds, purchased blankets, and other commodities, and commence their long walk to their kraals or location, in the warmer and more beautiful Zoutpansberg district, while some even cross the Limpopo River. The distance they travel is frequently over six hundred miles, and three or four hundred miles is a common journey. When on one of these long tramps they will often average eighteen miles daily, but a frequent rest for a few days at other kraals they may pass reduces the average of their daily pedestrian record. To see them toiling along with their heavy loads on head and back, frequently foot-sore and weary, but encouraged and sustained with the prospect of home once more, showed that these men had reached the elements of civilization. The labour question to them is not a matter of life-long servitude, and the few months spent working in the towns or delving in the mines is exchanged for an equal or far longer period of rustication among their own people. Some die on the road, especially in wet and cold weather, and we saw several who seemed to be thoroughly leg-weary and worn out. The money they have earned enables them to pay their yearly tax, but more particularly to find the purchase or "custom" money for another wife. Polygamy among these Kafirs is not necessarily a sensual institution. To women is deputed the whole manual work, both household and agricultural, and a wife will try and induce her husband to earn the means by which he can obtain another wife and thus lighten her own domestic duties. As is well known, oxen or money must be given to the father-in-law before his daughter can be obtained; but the heavy outlay thus incurred is an investment, and will be well repaid if the husband becomes the father of female children and so in turn becomes capitalist himself. In a savage or semi-savage community, women derive protection from such a custom. Female infanticide is unknown, the woman secures a safe and valued position in the tribe, and marriage thus having a financial value, any rampant immorality is discouraged and becomes an offence to the community.

Of course, some amount of immorality exists, as in the most puritanical districts at home, but at least a stand is made for the sanctity of marriage among these Kafirs by the prohibition of unmarried girls bearing children. It is very questionable whether these men lead more sensual lives with their few wives than they would do if they practised monogamy, and there are many occasions when the woman is avoided altogether, especially for some time after child-birth. At the stage of culture to which the Kafir has at present arrived, polygamy is a useful institution; it is a protection to the women, and an incentive to the industry and enterprise of the men. We are too apt to judge other social arrangements, especially when belonging to what we are pleased to call inferior races, by our own standard of civilization, which is often simply the subordination of the greatest good to the fewest number. Certainly, among these Magwamba Kafirs, woman only holds the place of a valued chattel (the women always kneel when handing anything to a man); but even then her lot is not worse, but probably better, than that of the well-abused drudge and slave of our own brickfields.

The Magwambas, or "knob-noses," so-called from having their noses originally ornamented with notches or scars, were the tribe or clan of the Bantu race with which I was principally thrown in contact. They entered Zoutpansberg about twenty years previously from the other side of the Limpopo, under the control or chieftainship of a Portuguese named Joan Albasini, and they still style themselves "path openers." They are mostly refugees from Umzila's country, since joined by other refugees from the surrounding districts, and are now the most orderly and law-abiding inhabitants of the Spelonken. At the death of Albasini they looked to the Transvaal Government as their head, and afterwards elected the government native commissioner as their chief, a proceeding they probably now regard in the same light as the early Jews did their insistance on having a king.

At the time of my visit to the Spelonken these

Magwamba Kafirs numbered, I was told, about twenty thousand. They do not live together in large numbers, but have small scattered kraals consisting of a few huts. A favourite dress of the men is a tiger-cat skin in front and often another one behind, and the women wear a short petticoat.

MAGWAMBA WOMAN CRUSHING MEAL.

There was a small kraal a little behind the store at which I stayed, from which lamentation had ascended for the last three weeks and still continued to resound across the wooded veld. The head man of this village, who started to work at Kimberley, had died on the road, and now funeral dances and loud songs of woe were still of frequent occurrence. An Induna who accompanied me to see these rites exhibited what is called the "scepticism of the better classes," and quietly remarked with a smile, as he handed me some Kafir beer, "it will not bring him back." All these men love strong liquor, and those who can obtain it show little moderation whilst the supply remains unfinished. Two Indunas visited the store daily, and patiently waited about during my visit, knowing that I had some whiskey, and by friendly smiles solicited the favour of being asked to take a drink. To look at these two men, there could be little doubt as to how they acquired their position. Good health, a stalwart and imposing appearance, the signs of mental capacity far beyond their fellows, a general air of good-natured cunning, and an absence of what might be called "morbid conscientiousness," made up the qualities that not only created success in a kraal, but with education would have made good men of business, who could have promoted Companies and held their own on a stock exchange. These are the attributes which for ever make impossible dreams as to the perfect "equality of man."

With these two Indunas we arranged the preliminaries for a great dance on the basis of my host providing an ox to be slaughtered and eaten on the occasion.

On the morning of the dance troops of Magwambas, ornamented with their most showy if scanty wearing-apparel and singing their songs or rather dirges, gathered in from all sides. Several Indunas were arrayed in war-like attire, and the whole scene reminded one of a public holiday at Hampstead or Riddlesdown at home, but without both the drunkenness and vulgarity. The only vulgar-looking Kafir was an individual in European costume, who had just returned from working at the diamond-fields. He was dressed in a suit of cords, his waistcoat was ornamented with three distinct brass watch-guards, he also possessed boots and necktie and wore a round hat; but, compared with his

[*To face* p. 102.

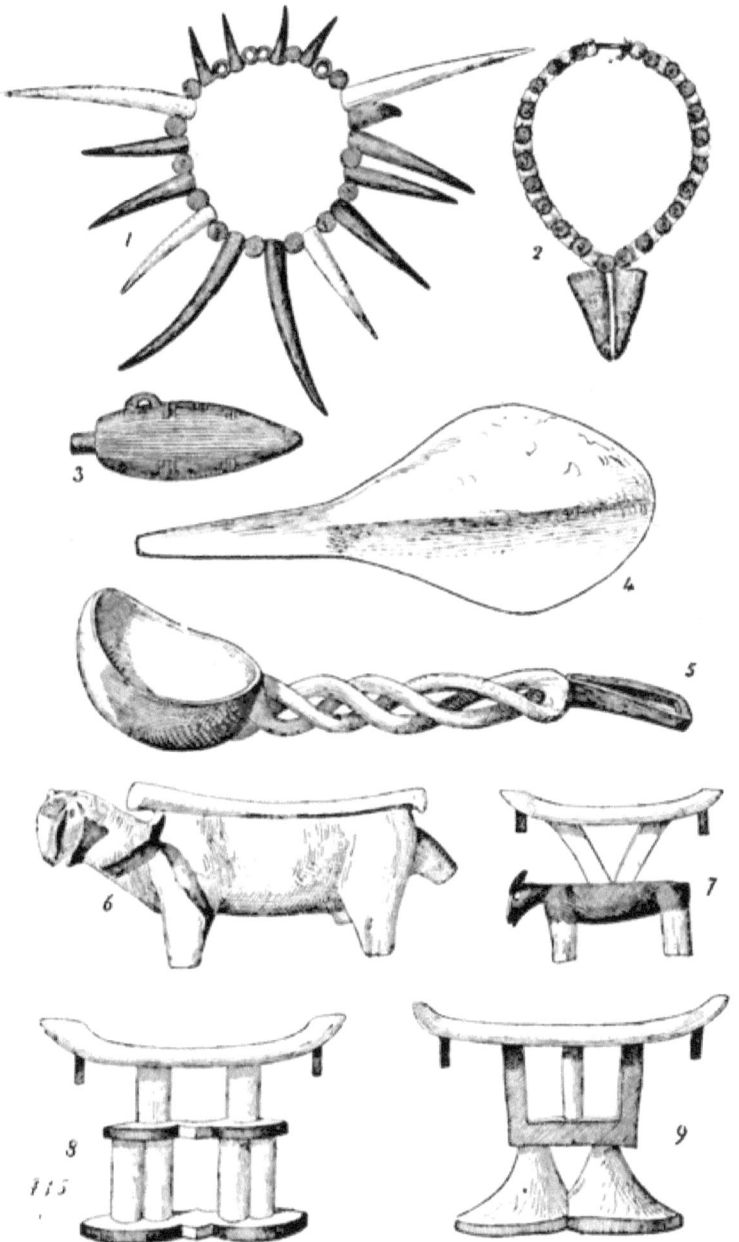

ARTS AND INDUSTRIES IN THE SPELONKEN.

1 & 2. Magwamba necklaces.   4. Mavenda pick or hoe.
3. Magwamba snuff-box.   5. Magwamba ladle.
6, 7, 8, 9. Magwamba head-rests.

less clothed but more artistically attired brethren, he looked like an East-end rough at home. Oh! nineteenth-century civilization, you have polluted the fairest spots with smoke and hideous erections, from which the factory bell tolls like a Newgate summons to the

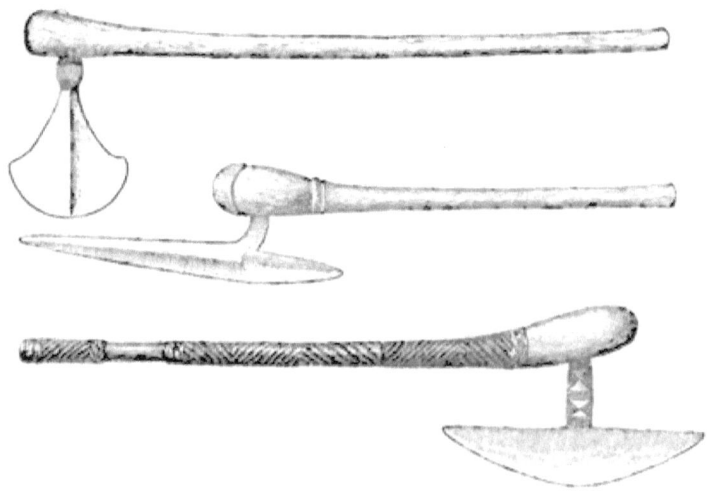

MAGWAMBA WAR-AXES.

condemned labourers; now in the Spelonken you send us such a vulgarized, if civilized, wretch as this! He dances not, he smiles not, he only looks on, but in a short time he will dispense with these hideous robes and once more dance and eat his mealies with his happy friends.

The dances might be described as of a "program" nature, and represented phases and events of Kafir life, such as "bearding a lion in his den," &c.; they commenced at about 11 A.M. and continued almost uninterruptedly till about 4 P.M. The men and boys formed a wide outer circle, inside which in two close phalanxes were the married women and unmarried girls. A Kafir really dances—he acts the dance, he enjoys it, he lives his part in it; to him a dance is a lamentation or a rejoicing, the glory of a fight or the story retold of a homely reminiscence: no wonder the labourer gladly tramps back from the large towns, where his existence is a compound of work and restriction, to the family life of the kraal. There freedom is combined with gaiety and excitement, wants are few, and their food simple and to hand. But the cry frequently heard from Europeans is that the government "should make the niggers work," and this by imposing heavy taxation. The advocates of this doctrine are often speculators, who believe that civilization consists in acquiring gold, and that the Kafir race should become one huge corps of miners to enable them to carry out the operation. For myself I often envied the simple wants and few troubles of these happy Magwambas.

During the dance, the unfortunate ox that was doomed "to make a Kafir holiday" stood a quiet spectator of the scene, but was assegaied as the afternoon progressed, and the process of flaying was commenced before the animal was quite dead. Kafirs have no regard for animal suffering; they carefully tend their oxen while alive, but when once it is decided to slaughter an animal, all consideration for the beast vanishes and the same individual can be as cruel a butcher as he was formerly a kind and attentive shepherd. The meat was quickly stripped from the carcass, numerous small fires were made, and the ox was soon a thing of the past. It is during such feasts that savage instincts are really seen, and we recognize that self-restraint and gentle manners after all are the true marks of civilization.

The authenticity of many travellers' accounts of the religious beliefs and origins of customs among so-called savage races have been long doubted, and on this journey I found the utmost difficulty in extracting any reliable or exact information from the Magwambas. I could only be told by one what was too often contradicted by another, and this, not because of their untruthfulness, but simply owing to our mutual ignorance of each other's meaning. Nor was it due to a want of knowledge of their language, as my host was a

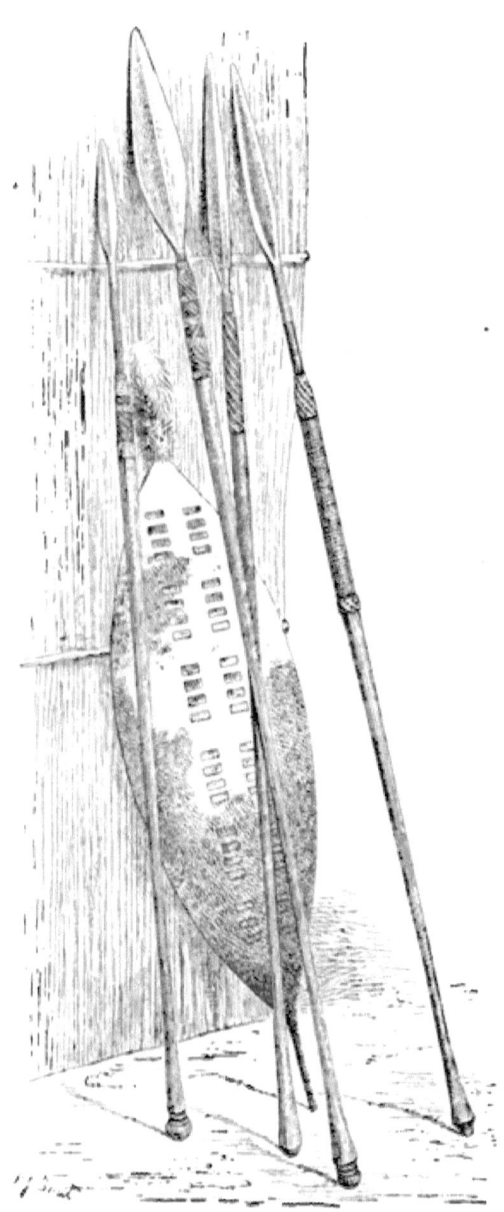

MAGWAMBA ASSEGAIS AND SHIELD.

thorough linguist, and, what was more, remained on the most friendly and trusted relations with them. One must live for some time with the Magwambas, *and as a Magwamba*, before any true insight can be obtained into their real speculative opinions, and then very few of them have clear notions on these points. It would be the same if a learned and anthropological Magwamba was possible, who should visit England and in a short time endeavour to study the origin and meaning of much theological and philosophical reasoning found in our midst. If he mixed only with our lower classes he would find little opinion at all; our middle classes would give him varied and often erroneous definitions; whilst among those of leisure he would find Galileos who cared for none of these things. So it is, in a more moderate degree, among native races, where are also found the totally ignorant, the thoroughly mistaken, and the supremely indifferent, as elsewhere.

The Magwambas are not the only tribe of Bantu Kafirs living in the Spelonken. The Makatese, originally fugitives from the Basuto and Bechuana countries and taking their name from the supposition that they were all subjects of Ma Ntatisi*, are now the most numerous in Zoutpansberg, and, under the chief Magato, are located on a long mountain range which exhibits one of the glories of the landscape. The Makatese, I was informed, now number upwards of thirty thousand.

The Mavenda Kafirs are a branch of the Makatese, and closely allied to the Basutos, and amongst these people iron-smelting and manufactured iron-work in a rough way is carried on. My friend arranged that I should witness the making of a "pick" or agricultural hoe, the principal article fabricated, and the head Mavenda sent me his pony on which to ride to his home on the summit of a hill, where I was received by himself and assistants under a thatched roof where the primitive forge was erected. The fire was soon

---

* G. McCall Theal, 'History of the Boers,' p. 63.

lighted, charcoal being used, and a small calabash containing iron (the ore procured from an iron mountain in the vicinity and previously smelted) was produced, the contents of which were thrown on the fire when sufficiently heated. When the metal was fused it was laid on a large block of stone and beaten into shape by another heavy stone wielded with great force by a stalwart and adept assistant, and it was interesting to watch how, with these rough implements, the pick slowly but surely grew into shape. It was taken from the forge by a rough pair of tongs held by the head man, who always whistled during the time he thus held it on the stone anvil, and his assistant with a grunt brought down his heavy weight on the exact spot indicated by his chief. During the whole time two men took it in turn to blow the bellows made of buck or goat skin, with the hollow horns of antelopes for the funnel, whilst several visitors squatted round and watched the operation. It was living in the iron age, and thought travelled back to the bygone times in human progress. These picks are greatly valued by Kafir agriculturists, always maintaining a value of about five shillings, and are greatly preferred to those made in Birmingham, which can be imported and sold for less money.

The manufacture of the pick forms thus a true native industry, and in this region is almost confined to the Mavendas, amongst whom, I was assured, there was a recognized compact that none should be sold under a certain price. The Mavendas by their industrial arts are thus more advanced in material progress than the Magwambas, with whom they live in contact, though the Magwamba women always wear a petticoat, and the female Mavendas have simply the ordinary waist-bandage. But though much less clothed, the Mavenda women are better-looking and exhibit the signs of more intellect than the Magwambas possess. Material progress and clothing certainly do not always go together.

I considerably added to my natural history collection during the ten days I spent at the Spelonken, awaiting

Native Iron-smelting.

the wagon for my return journey, and in this I was greatly assisted by the Magwamba boys, who, on finding that there was really a market, set thoroughly to work in procuring specimens. Birds were mostly brought alive, as the lads were adepts at trapping, or when killed they were generally in perfect condition, as the blunted wooden arrow-head was used. At first some of the men would bring a small bird pierced by a bullet shot from an old "Brown Bess"; but they soon knew the requirements better, and a good ornithological collection could have been obtained had I possessed leisure to remain longer on the spot. The great trouble was to prevent them bringing the same thing over and over again, and to make them understand that insects were valueless when crushed; but they really experienced pleasure in trapping and shooting birds, and would attentively watch the process of skinning. As the lads brought in my prizes, I recalled the same arrangement made years before with the Nicobarians in the Bay of Bengal and the Malays of Province Wellesley.

Animal life was, however, scarce, the dry season had just commenced and birds had generally left the neighbourhood. The only predatory beast was the Jackal (*Canis mesomelas*), whose shrill cries or screams had broken our rest and disturbed the deep stillness of the night as we journeyed up in the wagon. On our arrival at the store we heard that these animals had been prowling around and had dragged away a dried hide a few nights previously. Buck were very scarce, one species only, the Duyker (*Cephalolophus grimmii*), being obtained during my stay. No quantity of big game can now exist near a Kafir location since the introduction of firearms, and the natives have learned to use a gun with much greater precision than in their early fights with the Boers, when they frequently shut both eyes before firing.

The dry veld now no longer contained its rich variety and myriad numbers of orthopterous insects, and this, I believe, was the cause of the almost utter absence of birds in the spots where they were previously so abundant. In the Transvaal I found that almost all birds fed on this rich banquet of the rainy season, and I have even seen the crops of Kestrels (*Cerchneis tinnunculoides* and *amurensis*) crammed with the remains of these insects; the Short-eared Owl (*Asio capensis*) also feeds on large Coleoptera, the crop of one specimen I procured containing nothing else. As soon as the dry season recommences there is an absolute dearth of insect-life on the veld, and birds must then seek other areas in quest of food. The most showy bird in the Spelonken was the Roller (*Coracias caudata*), and the curious cry of the Grey Plantain-eater (*Schizorhis concolor*) was generally to be heard when one rambled among the trees; whilst in Francolins, *Francolinus subtorquatus* and *F. gariepensis* replaced the *Francolinus levaillantii* which I had recently found so plentiful in Pretoria. Here also I observed and obtained the great Jackal Buzzard (*Buteo jakal*), which I never met with in the Pretoria district.

The best entomological spot found in Zoutpansberg was on the banks of the Dwaas River near the ford which forms part of the high road; on the damp sandy banks hovered clusters of small yellow butterflies (*Terias brigitta* and *T. zoë*), like constellations of primrose-blooms, and in the same spots the dragonfly (*Trithemis sanguinolenta*) literally swarmed; besides these species I procured, during a half-hour's stay, the pretty *Teracolus subfasciatus*, besides several other species of the same genus, and on the wing captured different species of *Buprestidæ* and *Longicornia*. As this was at the end of the summer, it should prove a good locality at the right season for a travelling collector. I could have pleasantly passed the day on these wooded and sandy banks, but the oxen were once more inspanned and my friend was anxious to resume his journey home. Species of *Teracolus* abounded all along the road, and I often walked behind the wagon net in hand with the best results; it was thus that I captured the only specimen of *Teracolus vesta* I found in the Transvaal.

Zoutpansberg is one of the richest districts of the Transvaal, if not the very richest, so far as fertility of soil is concerned; its auriferous deposits are highly spoken of; its scenery is in many places superb and in strong contrast to the melancholy monotony of the high veld. To leave Pretoria and in two or three days reach the natural beauties of Zoutpansberg, after necessarily traversing the pleasant Waterberg district, is like exchanging a wilderness for fairyland. That high tableland of treeless veld, with its everlasting monotony of plain and kopje, is fit abode for the quiet and unimaginative Boer; its very sameness reminds him, or, rather, appeals to his fancy, of the plains of Palestine, of which he reads so much and understands so little; solitude not nature appeals to his mind, and Wordsworth in these worthy folk would have found a people who had given their hearts away from nature without the excuse of the world being too much with them. But when we descend to the lower lands of Zoutpansberg, with its warmer air, its rich vegetation, and its happy Kafir population, our touch with Nature seems to be once more resumed. However, Zoutpansberg is not alone destined for the dreams of a Rousseau, it may yet prove the gem of the Transvaal. Give a rail connecting it with Pretoria and from thence to the sea, and this fertile land would produce the richest farms on the face of the globe. What incentive is there now to struggle for an agricultural produce that could find no market? this long and costly transport would prove the ruin of the farmer who cultivated this life-giving land. Take maize alone and compare its value in Zoutpansberg with its price in Pretoria, and still the much lower figure is more profitable to the grower than the higher obtainable in the capital, for the cost of carriage would entail a loss, and the time employed for the same would prove the destruction of all fresh goods that demand early consumption. A rail would also develop its mining capacities, and had these lines been earlier constructed Pretoria might have become the terminus for Central Africa.

As I returned the dry season proclaimed its advent by the frequency of grass-fires, and the few residents one met affirmed that the rains were over for the season; so certain were they on this score that my wagon was not even provided with its sail-covering in case of wet, an omission that might have caused much discomfort, as a storm went before us to Pietersburg and exceedingly heavy rain fell there on May 2nd, the day before we arrived. At Pietersburg we met men going up to and coming down from Mashonaland; and though much doubt still exists, we shall see whether British enterprise in that new Protectorate is not as capable of producing a country, from a "geographical expression," as successfully as other influences have created the Transvaal, thanks to its auriferous deposits and its attendant European settlers.

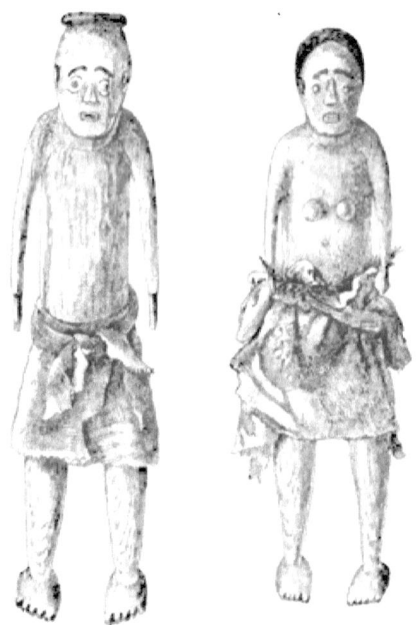

MAGWAMBA CARVINGS.

(115)

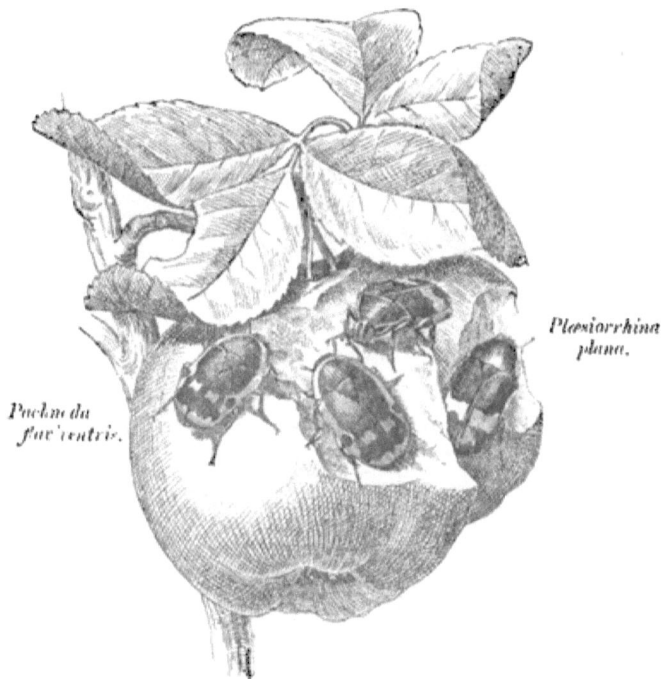

Pachnoda fasciventris.

Plæsiorrhina plana.

APPLE-DESTROYERS IN NATAL.

CHAPTER VII.

A JOURNEY TO DURBAN.

Acacia mollissima.—Heavy cost on imports to the Transvaal.—Johannesburg and its Hotels.—Heidelberg.—A Priest of Islam.—Across the Ingogo heights to Newcastle.—Durban.—Colonel Bowker.—Best collecting-grounds around Durban.—Flowers, fruit, and insects.—Peculiarities in railway construction.—Model Natal farms.—Insect-pests to gardens.—Difficulties in coaching after heavy rains.—The store- and canteen-keeper of the veld.

IN December 1890 I journeyed to Natal for the purposes of visiting the farms where the Wattle (*Acacia mollissima*) was cultivated, from which is stripped

I 2

the "Mimosa-bark," that now supplies a large quantity of tanning-material for export to England. I may here at once state, and the fact will explain the difficulties and expense of transport to the Transvaal, that I found this bark could be absolutely delivered at £2 per ton less in London than at Pretoria. The cultivation of these "Wattles" is largely on the increase, and will considerably add to the exports from Durban.

Christmas had passed without pleasure, for, even stripped of the pagan accessories of holly and mistletoe, it will always be to Englishmen a time of family reunion, and my thoughts were with my family away in snow-covered Surrey. On "boxing-day" I left by the coach for Johannesburg, and once more began to retrace my steps towards the sea. It soon commenced to rain, and we subsequently drove through a white mist or damp fog, such as I had not seen since leaving home, and which seemed little in keeping with what one anticipates in South-east Africa.

Johannesburg, which we reached about 7 P.M., is the veritable Chicago of South Africa. The Rand is high, healthy, and cool, and the atmosphere quite invigorating after the close and still air of sheltered Pretoria. The surrounding country looks bare and desolate in the extreme, there are scarcely any trees to be seen, there is nothing picturesque, but there is Johannesburg and the site of the finest gold-producing reef in the country. It is here that the real pulse of the Transvaal is felt, though the heart may beat at Pretoria. Young trees are being planted in considerable numbers, and by the time these have grown and added sylvan beauty to the spot, may commercial prosperity also have returned to a town that holds so many of our countrymen and contains so much capital belonging to English investors. Gold is the main strength of the Transvaal, but its quest by unscrupulous company promoters has been its curse.

It was a great relief at Johannesburg to once more stay in a comfortable hotel, especially with single-bedded rooms. To occupy a double-bedded room without having seen the other guest, who may turn in very late, and perhaps not sober, is what I once or twice experienced during my stay in the country, but which I gladly relinquished when possible. The Grand National Hotel was built at the time of the greatest prosperity of Johannesburg, and the cashier assured me that the takings then averaged £200 daily; but at the time of my visit the receipts only amounted to about a fourth of that sum. Some of the smaller hotels were virtually closed, and fires had become somewhat frequent. I remember years ago travelling with an American who praised in no half-hearted way his native Chicago. Some considerable conflagrations having recently occurred there, I ventured to remark that fires sometimes took place. That is nothing, replied my companion: "when a business man in Chicago is going to fail, he burns his place down."

I left Johannesburg at 5 A.M. on a fine Sunday morning for the long coach-drive to Newcastle. The journey was scarcely different to what I had formerly experienced, save that the coach was less crowded and the veld was now green in its summer dress, whereas it was in winter brown when I crossed before. But the sky was now clouded, frequent showers of rain occurred, and one missed the lovely warm umber tints of the veld and hills as seen under the clear winter sky.

We reached Heidelberg about 10 A.M., a small and early established town, but, like the rest, suffering from the present depression. It has a considerable "coolie" or Indian population, and a priest of Islam, who had been travelling through the South-African diocese, joined the coach. He was certainly one of the most handsome men I had ever met. Tall and of graceful stature, he possessed a perfectly formed and chiselled mouth, such as one seldom sees in man, but is found principally in women of the classical type of beauty; and with an aquiline nose was also combined the dark soft spiritual eyes which mark the true type of the visionary priest of all creeds. This man was evidently of good birth, as proved by the ease, confidence, and repose of his manners; he was seen into the coach with great respect by his native friends, and was in like manner received when he arrived at Standerton, for this peripatetic theologian was evidently entertained like a minister visiting a conference at home. At one of the stages, where we changed horses, there was a Boer's house, where tea was supplied to travellers at sixpence per cup. I and my other British travelling companion entered a small room to partake of this soothing beverage; the Mussulman followed, but was indignantly warned away by the Boer woman to take his tea outside. The good-natured ease and polish expressed in a wave of the hand, by which he declined this form of entertainment, was in strong contrast to the dull features marked by stolid ignorance or stupidity belonging to the female dweller on the plains, in whose eyes this man was simply a common "coolie," no more than a Kafir entitled to enter her humble abode, or associate with her white customers. It was the meeting of ignorance with education, but with power in the hands of the first. The roads were very heavy, and the hard iron paths, over which we had previously travelled on our way out, were now often replaced by miles of soft mud, through which the coach progressed with the greatest difficulty. We reached Standerton about 8 P.M. to dine and sleep, with instructions to be ready to start at 3 A.M. the next morning. This was carried out punctually, and as the proprietor of this "Hotel" only provided early coffee, when there were not too few passengers to make it financially worth his while, and, as there were only two on this occasion, we started at break of dawn, and drove 16 miles before reaching a small and lone canteen then surrounded by a sea of mud. Here the welcome coffee was obtained. By noon we arrived at the confines of the Transvaal, entered Natal, and were once more under the old flag.

We now changed coaches and started for Newcastle, traversing again the spot where Briton and Boer met in deadly and unnecessary conflict. The hill at Laings Nek was in a very bad condition, owing to the late rains, and the oxen drawing the transport wagons were terribly distressed as they drew their heavy loads up the steep ascent and through the deep mire. From this part to near Newcastle the road was one of the worst I had ever travelled over. We had exchanged the ponderous coach for a light kind of wagonette, which was better able to traverse the yielding soil; heavy rain descended and came through the canvas roof and side coverings of the vehicle; water poured down the steep hill-roads in rivulets, and the scene and surroundings were desolate in the extreme, especially when we crossed the Ingogo heights, where monument and cross denote the burial-place of so many British soldiers.

Our driver was a Cape boy, our conductor a half-bred Indian, whose father, he told me, had been an Englishman. Both exhibited an inclination to make merry of England and her soldiers on a basis of Boer supremacy. As a delicate piece of sarcasm the driver at length asked me if we grew pine-apples in England. Certainly, I replied, in glass houses at home, and plentifully in the open air in that part of Britain called Natal. But you would not call me an Englishman? he asked, in startled surprise. Certainly, I replied, if you were born and are living under the British flag, under British law, and prepared to maintain British rule. Ah! but, he remarked, all Englishmen don't say that, most of them call Natal "Kafir-land." I cannot help that, I responded, I call Natal England as much as I do Scotland, and one day, I hope, Ireland.

We reached Newcastle about 6.30. This town is rapidly becoming a prosperous one; it possesses abundance of coal in the neighbourhood, but the Transvaal Government have placed a prohibitive duty on that article being imported into the Republic, which is thus prevented from becoming a customer. Probably, however, Newcastle has reached its zenith, and the railway will not only pass it by, but carry a considerable portion of its trade to the terminus at the frontier.

Since my visit, the railway, in April of this year 1891, has been completed and opened to Charlestown, thus bringing the line to the Transvaal frontier and under the shadow of Majuba Hill. Thus far it has come from the sea at Durban; its continuance depends upon the sanction of the Boer Government. The progress of this line has been slow but continuous. The first instalment was made in 1860, with a short section connecting the point with Durban. It was not till 1873 that a further move was made, when powers were obtained for pushing on the line to the Drakensberg. Pinetown was reached in 1878, and the capital (Maritzburg) two years later; Howick, 1884; Ladysmith, 1886; Elands Laagte, 1888; Biggarsberg, 1889; Newcastle, 1890; and now Charlestown and the frontier, 1891. The great engineering feat of this last extension is the tunnel through Laings Nek, which was bored through a hill 3200 feet wide (the actual tunnel is 2213 ft. 6 in. long), consisting of the hardest indurated shale, with the addition of three dykes of whinstone. It was completed in October of this year 1891, after having occupied an average of 430 coloured and 60 white men upwards of two years in its construction. About 85,000 tons of slate and whinstone have been excavated, and upwards of 8000 tons of masonry have been required for the purposes of lining; there have also been used over 40,000 lbs. of dynamite, 4000 lbs. of powder, 70,000 yards of fuse, and 50,000 detonators in the necessary blasting-operations. Thus Laings Nek may now be associated with a monument of our colonial enterprise, and its painful military memories be forgotten. Already, when I passed down, the neighbourhood of Charlestown was being covered with the iron-roofed huts of the advance guard of commerce, and soon many of the spots celebrated only for a useless carnage will be almost obliterated by the dwellings of a trading community.

The train left Newcastle at 7.15 P.M., and thus the night was passed in traversing a region which I had seen by day on my journey out. Political divisions do not alter the physical aspects of a region, and after passing the grand mountainous scenery between Charlestown and Newcastle, the country once more resumes the flat and uninteresting appearance of the Transvaalian veld. This is gradually transformed after leaving Ladysmith, but does not altogether improve till we approach the neighbourhood of Maritzburg. Then lovely valleys, grand mountain-gorges, stretches of hills rolling far away and fading on the sky-line, beautiful verdure, and (what was even more to the wearied Transvaal eyes) forest appeared, interspersed occasionally by small rivers or spruits. Halfway between Maritzburg and Durban we are reminded of greater warmth and another aspect of vegetation. Tree-ferns and bamboos are now seen, and lower down fields of bananas and pine-apples, with patches of sugar-cane, recalled old days lived in the tropics.

It was raining at Durban when I arrived, and there was a mist over the sea; but what a pleasure the sight of the ocean is after living on the dry and almost waterless tableland of the Transvaal! Of course the usual "Currie" and "Union" steamers were seen at anchor, and these, with other steam lines, have now as effectually superseded the sailing passenger-vessels that formerly journeyed to Natal as though they had rammed and sank them.

The Museum, which occupies a well-lighted and lofty apartment above the Town Hall, is in course of evolution. It is poor in mammals, but is beginning to obtain a good collection of birds, well set up and in cases that contain much available room for additions. In insects the strongest element is butterflies, a fine collection properly arranged and named, as might be expected in a town possessing as resident such an old lepidopterist as Colonel Bowker. I was glad to meet the Colonel, the best field entomologist in South Africa, who has invented a net he wears on his hat like a puggaree, and which is ready to be affixed to the stick he carries in his hand for instant use when a desired specimen is seen, whilst an original pocket collecting-box has also been devised by this active lepidopterological brain. I spent New Year's day with him at his bungalow at Malvern, a suburb of Durban, a lovely spot, embracing on one side an extensive sea-view, and behind a vast extent of undulating scenery such as Natal can so lavishly exhibit. It is such a subtropical spot as a naturalist might choose in which to happily live and cheerfully die.

Col. Bowker is now endeavouring, by cultivating the old Natal plants and flowers, to prevent many of them being practically relegated to oblivion, and I was particularly pleased to see, entangling a bushy tree, an old friend of my greenhouse at home, the Jamaica Passion-flower (*Passiflora quadrangularis*).

Durban holds high holiday at the advent of the new year, for here Scottish blood flows thick, and the old days of Scotland are ne'er forgot. I could thus for some days pursue none of the business for which I had visited the port, and was able to pay some attention to its entomological attractions. As soon as I arrived I saw I was in a rich spot of insect-life, and one possessing a different facies to that of the Transvaal. Around the trees in the town flew a handsome moth (*Egybolia vaillantina*), whilst at evening, in the smoking-room at the hotel facing the sea, the fine Saturniid moth *Urota sinope* paid occasional visits attracted by the light, accompanied by lamellicorn beetles of the genus *Adoretus*, and other insects. A stroll in front of the hotel before breakfast resulted in the capture of the pretty longicorn beetle *Rhaphidopsis zonaria*, the morning after my arrival, whilst butterflies swarmed over the scrub that covers the back beach.

There are three good spots for the collector in Durban, and each easily approached. The first and probably the best is the "Bluff," the headland at the harbour mouth, and on which the lighthouse stands. It is backed by an extensively wooded district, and a somewhat representative collection of Natal insects might be made during a season's work at the spot. The second best-ground is about the Berea, the high land overlooking the town, where the principal residences are also found ; here the entomologist should seek the site of the old windmill, approached from the Toll Bar on

the Tram Line. The third hunting-ground is in the
"Wood" or "Bush" that runs continuous to the beach,
and to one whose time and opportunities were limited its
proximity to the hotel rendered it available for an otherwise busy man. Here butterflies haunt the narrow
paths, cicadas (at the time of my visit *Platypleura
punctigera* was the dominant species) utter their shrill
cries, and beetles are probably abundant at the commencement of the wet summer season. I found here,
as at the Transvaal, that after rains, when the leaves
were damp, more beetles could be found on them than
in dry weather, when Coleoptera are to a greater extent
on the wing.

It was a delightful sensation to be roaming in these
thickly wooded glades, though in Durban summer heat.
Butterflies abounded, of which the most common were
Acræa natalica and *Planema esebria*, high up amongst
the trees flew *Salamis anacardii*, ever and anon down
the narrow paths came the sulphur-and-red *Eronia leda*;
Papilio morania and *P. demoleus* were not uncommon,
whilst *Teracolus* of many species enlivened the scene.
In moths the gaily-marked *Euchromia africana*, by the
rapid vibration of its wings in flight, would cause considerable doubt as to what insect was observed, and,
till I became acquainted with its peculiar habits, I
frequently mistook it for a species belonging to another
order of insects; the modest *Leptosoma apicalis* was
found in shady nooks, and *Aphelia apolinaris* took long
and high flight in the clear light of noon-day. The
showy Neuropteron *Palpopleura lucia* flitted about,
and on blooming plants I not only found the handsome
beetle *Popillia bipunctata*, but a variety of Cetoniids
such as *Coptomia umbrosa*, *Elaphinis irrorata* and *latecostata*, *Trichostetha placida*, and the curious Telephorid
Lycus bremei.

For a naturalist, especially an entomologist, intending
to study the fauna of South-eastern Africa, Durban is
the best introduction to the country. A month spent
at this port collecting and observing would give a
thorough introduction to the southern portion of the

Ethiopian fauna. Subtropical Durban could thus become a tropical training-ground for the exploring naturalist, who would be able to develop that simplicity in requirements and acquire that amplitude and method in observation which are so often more laboriously learned at the cost of missed opportunities when he reaches the interior. There is a lore in collecting natural objects which can only be acquired by practice, for until the habits and haunts of animals are understood they will not be searched for in the right spots, and necessarily will not therefore be found. A traveller often passes over a rich and unexplored zoological region which he only samples through having had no preparatory training as might for Southern Africa be obtained at Durban. But, of far more importance, the power of observation is quickened by an early appreciation of what and how to observe, so that the capture of an animal will soon become of less importance than a knowledge of its relation to its environment. I could not help contrasting the different mental conceptions which dominated me when collecting in the Malay Peninsula twenty-two years previously and those which now occupied my mind in a similar quest at Durban. Then almost the sole aim was the discovery of new species; now the constant wish was to make some small discovery to add to the ever-increasing knowledge of how animals derived their present shape and coloration in the struggle for existence. These pleasant Durban glades, where insect-life so freely exhibited itself, were now no longer only emporiums to supply museum drawers with specimens, but were full of nature's records of the past—like hieroglyphic writings, but unlike them, most at present we cannot read. It was now the cult of Darwin that seemed wafted in the air, and I felt like an eclectic Pagan finding a shrine to philosophy amidst these African groves.

It was on a Christmas day that Vasco da Gama reached and named Natal, at the height of summer and amidst the glories of a vegetation as I now saw it. Although four hundred years have elapsed since that discovery, Natal has only been colonized in quite recent times, and its flora, save by introduced plants, has had insufficient time to be radically altered. The gardens were gay at the time of my visit with the flowers of several varieties of *Hibiscus*, aloes exhibited their huge flowering-spikes, and lovely creepers in full bloom were quite common. The fruit-market rejoiced with pineapples, mangoes, bananas, and other vegetable products of a subtropical nature, which add a charm to Durban, and, apart from its summer heat, I would more willingly live and die at that port than in any other part of Southern Africa.

The beach in front of the hotel at which I stayed had, however, other characteristics beside its beauty. One suicide and two dead bodies washed up during my stay of five days was rather a ghastly, though, I believe, unusual spectacle. One body was described to me by a "morgue" enthusiast as particularly curious from the fact that only the face and boots were perfect, and he seemed somewhat chagrined that I did not allow him to guide me to this gruesome sight before lunch, for which he said "there was just time."

I left Durban by the morning train for the north, intending to visit the bark farms of the interior. Perhaps few railways exhibit more singular freaks of construction and engineering skill than this Natal line. When mountains cannot be avoided, the rails run round them in serpentine arrangement, and to avoid these elevations, when possible, the line takes such a devious course as to frequently give the impression that one is returning to the spot only just previously left, while the curves are so sudden that you can often see both ends of the train at the same time. A story is told of an engine-driver who pulled up in obedience to the danger-light belonging to the rear guard's break of his own train, which, in the intricacies of the curves, had become placed in front of him.

Owing to the steep gradients, a single line of rails, and the number of small stations, progress is very slow, and my return journey did not average much more than twelve miles per hour. I was much mistaken in the character of a fellow passenger who joined the train at Ladysmith. On asking if he smoked, I received such a determined answer in the negative, with an assurance that he had never done so, that I took him for somewhat of a Puritan. He soon, however, produced a bottle of whiskey, which, by assiduity and perseverance, he quickly emptied and then lay full length and speechless on the seat before me.

I broke my journey through Natal at Richmond Road, and had the pleasure of being entertained at a home of comfort on a model African farm. Here was a well-built residence furnished with taste, containing all the comforts of a home, and a library sufficient to prove that a farmer can be a gentleman and cultivate his mind as well as the soil. What a contrast to the Boer farmers of the Transvaal! I do not speak disparagingly, but comparatively. Men cannot for ever trek on into the wilds and live solitary lives with their families without losing most qualities of domestic refinement, even though acquiring personal independence. In pursuit of game or on a hunting expedition let me be allowed to accompany the Boer and share his wagon; but the tie snaps when the time comes for the pleasures of personal intercourse and home life.

In the fine garden attached to this Natal residence I was shown the difficulties attending the labours of the horticulturist owing to the ravages of injurious insects. The roses were literally covered and devastated by a Cantharid beetle (*Mylabris transversalis*) and his apples were being completely eaten by two other beetles belonging to the family Cetoniidæ (*Plasiorrhina plana* and *Pachnoda flaviventris*). His principal enemies which occasioned his heaviest losses were the ticks (*Acarida*), which attacked his live-stock with the most disastrous results; clearly there is room for a state-paid economic naturalist in Natal. I was interested to learn that even in this colony, as in the Transvaal, material and industrial progress had been much retarded by the presence of the financial agent and company-promoter, of whom I was assured there were quite a colony in Maritzburg, and who my host described as " Hebrew Lilies, who toiled not neither did they spin."

I left his house to catch the night train, driven in a

Mylabris transversalis on Rose.

Cape cart drawn by a pair of spirited horses guided by a native boy. The night was pitch dark, the roads bad, with a river to drive through, and yet we went at full speed without a single sign of hesitation on the part of the boy who held the reins.

As the train sped along, and for thirty miles before we reached Newcastle, we constantly disturbed small flocks of the South-African Kestrel (*Cerchneis rupicola*). These birds were usually found two or three together and often on the ant-hills which bordered the line, taking flight as the train approached; but I saw very few birds during this journey, and a fine pair of Paauw (*Otis kori*), walking on the open veld near the Ingogo heights, were the most interesting of our ornithological observations.

At Newcastle I once more joined the coach for Volksrust, the first stage in Transvaal territory, and found as a travelling companion an Englishman who had been through the Boer war, and one of whose duties was now to see that the graves on the summit of Majuba were properly preserved. Again I listened to a truthful account by an eye-witness of the disaster—for I had previously travelled with the war-correspondent of a London daily paper, and also of a carrier of despatches during the war,—and again was the problem intensified as to the cause of all our disasters. He told me he had guided many travelling British officers up the mountain since the war and they always returned dispirited and perplexed. The disaster of Majuba has yet received no rational explanation, and it is said of General Joubert, that when the subject is mentioned he usually raises his hat and says the God of Battles fought for the Boers on that day. I have been on the spot three times, I have conversed on the subject with three eye-witnesses, I have heard a score of different theories about the fight from men of different nationalities whom I have met in the Transvaal, and I confess I do not understand it, and thought I knew more about it before I left England.

Twelve days' heavy and continuous rain is sufficient to incommode the inner communications of any country; but in a land like the Transvaal, where the rivers have few or no bridges and the roads are self-worn by coaches and wagons across an endless veld, the way becomes stopped and communication almost ceases. On reaching Volksrust in the rain and dark after a terrible journey across the flooded heights, we found no coach had been able to pass for three days, and the passengers had thus accumulated in the small wayside rest that passes by the name of hotel. The cause of the delay was the swollen and impassable condition of the spruit we had to cross a short distance further on, which had been daily approached but never attempted. We retired to rest—four beds in the room,—the latch of the door out of order and the rain pouring outside, and rose again at 3 A.M. to once more try and force the passage of the stream. When we reached it daylight had well broken, but there were ominous murmurs that the water was as "high as ever." A Kafir was sent across as a preliminary plummet: the stream reached his shoulders at the deepest part and carried him off his feet where the current ran strongest. It was, however, decided to risk the passage, by unharnessing the horses and letting the coach be pulled through by our Kafir friends, who now mustered somewhat strongly. Several of the passengers undressed and preferred swimming the stream to the danger of being overturned and washed away in the coach; but I chose the latter alternative with a prospect of being able to keep warm and comfortable; nor was I mistaken, for the coach, after threatening for about sixty seconds in the rush of the stream to end its present career of usefulness, eventually passed soberly through and gained the other side. Our passengers now again dressed, the horses were swam through, and after having chased a runaway, that broke loose and enjoyed his liberty for half an hour, we resumed our journey.

The Waterfal river was reached about sunset, and in place of the small and fordable stream we had crossed without difficulty some twelve days previously, there now flowed a wide and, in the centre, deep current not to be ventured by coach or horse. We crossed in a punt, and leaving our coach, transferred ourselves and baggage to a heavy wagonette that was waiting for us on the other side, and then travelled in the dark—for there was no moon—till about 10 P.M., when a stage was reached where we could sleep for the night. Most, if not all, these sleeping and feeding stages are kept by men who combine the trade of store- and canteen-keeper with that of "hotel proprietor." These stores are small shops, which, however, contain everything that the Boer customers (for the farmers ride long distances to these *dépôts*) require, from grocery to saddlery, including cheesemongery, drapery, bread and spirits, boots and crockery, ironmongery and tobacco, in fact all the requirements of a somewhat rude existence may be purchased at these unpretending shanties. Strange to say, the prices do not appear to be very high, and taking into consideration that all the stock is bought at second-hand, with a heavy transport added, the profits must be only moderate. However, these merchants are frugal, and their household expenses are necessarily light, so a moderate profit with a small turnover is yet something that produces a balance at the end of the year; but the life must be dreary and monotonous, the arrival of the coach the only communication with the outside world, and a few Boer customers the only other visitants, to break the ever present silence of the vast surrounding veld. Before daybreak we had again driven on, crossed the next river on a small raft floated by casks—a driver of another coach was drowned here later in the same day—and eventually reached Johannesburg at 9 P.M., instead of 1 P.M. the proper time. As we drove through the streets the canteens seemed to be the only places open, and I was told that at the time of the boom there were 280 of these establishments in the town; now the trade has decreased, there is little money for a carouse, and bad spirits are more difficult to sell.

The details of this return journey will give some conception of how the development of the Transvaal is retarded by the want of railway communication with Natal and the sea. This is the road to be pursued and these the rivers to be driven through by the ox-wagons which bring up the supplies of the country. Along this route travels the coach with the mails, and all the other main arteries of the Transvaal are of a similar nature and in a like condition. How can commercial prosperity be established under such conditions? But, on the other hand, did the farmers who trekked from the south and acquired this country desire the establishment of a *commercial* community and a network of railway communication? The answer is clearly in the negative; it is the "Uitlander" who has been the pioneer and is still the support of commerce in the Transvaal; and though the intelligence of the Boer community sees clearly that it is in its exports and imports that the prosperity of the country will largely depend, one can still sympathize with the majority of the farmers, who love the silence of their farms with the quiet existence and few wants, and enjoy that independence of life and character which is not usually attendant on trade, with all its other advantages.

KAFIR SHEPHERD.

CHAPTER VIII.

THE MEN OF PRETORIA.

The inhabitants of Pretoria.—Auriferous wealth alone the present cause of Transvaal development.—Uneducated condition of the Boers.—Liquor traffic with the Kafirs.—The British colonist in the Transvaal.—The Hebrew in Pretoria.—Commercial morality.—The name of Mr. Gladstone execrated in the Transvaal.—The Kafir and his value as a labourer.—Sanitary condition of Pretoria.—Vital statistics.—Aftereffects of the boom.—Attachment of Colonists to their adopted country.

Who are the men of Pretoria? We have already spoken of the Boers, who are farmers and burghers, and not what are understood as merchants and citizens; and in dealing with the population of the capital of the Transvaal—a question more for the sociologist than the anthropologist—the Boer may be treated as a non-resident factor altogether. Although Pretoria is the seat of government which is Boer, and the residence of the President who is a Boer, and the principal church in the principal square of the town is a Boer church, yet the Boer is still an emigrant, country-born Dutch farmer from the colony, who has to rely for judges, clergy, and civic administrators on the educated Dutch of the colony, or, what is worse, imported Hollanders who have neither the independence of the South-African spirit nor the necessary knowledge of local customs and institutions. The Boer is a farmer pure and simple; the commercial prosperity of the Transvaal is a thing which he has not created, and which it is just possible he does not desire. The auriferous quartz of the Republic, worked by emigrants from all parts of the world, but principally by British, have filled the coffers of the Boer Treasury, made the large towns of the Transvaal, and brought the country to a position from which it must either advance or retire. Had it not been for the mineral wealth of the Republic, its exports must have principally consisted of wool, hides, and skins, and it would have remained more of a large farming community than an industrial and political organization. But the dream of the early *voortrekkers* for a modern Palestine has not been realized; in a land of mineral wealth it could not be; they have many of them acquired wealth, but it must in the end prove their political extinction; their only chance of permanency is to form one of the units or component parts of a United South-African Confederacy, which in a hundred years they might influence, but less govern than Cromwell's puritans do the England of to-day. The Boer has only one chance to prevent his relegation to oblivion in a country of which he literally possessed himself, and which he secured only by rough living and hard fighting. That last chance is to immediately have his children properly educated; and, from what I observed, that course is not likely to be pursued. What he won by the gun will be lost by the pen, and the trade and *actual* government of the country will be in the hands of the colonists from whom he trekked, the British whom he not unnaturally dislikes, and the Hollander, who in his heart he distrusts and hates.

The present inhabitants of Pretoria, ignoring the few Boers for the reasons just given, are Colonials—that is, descendants of colonists who have settled near the Cape or at Natal and even Australians,—the Briton, the Hollander, the German, and Jews of various nationalities. Some of the largest stores belong to old British Colonists, who have come up to the Transvaal and now form a nucleus of the most reliable residents in the state. Many of these men are wealthy, their property is inseparably connected with the Transvaal, and having a large stake in the country, they do all they can to preserve its integrity, to develop its resources, and to improve the social condition of the towns in which they live. In the earlier days, before the influx of the mining migrants, their trade was principally carried on with Boers; and though they now do a larger trade with the different nationalities who have made the Transvaal their home, they have not forgotten their earlier and still constant customers, and are true in their allegiance to the Republic. Fortunately some of these merchants do also a large Kafir trade, and the aborigines thus procure advocates whose interest it is to see that they are neither driven from the country nor prevented from earning a just wage, some of which must find its way to the store. Though undoubtedly large quantities of alcohol do pass from their hands and through their agents to the native races, to the utter demoralization and physical deterioration of the Kafir, who *cannot* drink in moderation when liquor is to be procured, the injury thus done is somewhat compensated by the interest these traders bestow in seeing that the Kafirs are not unduly oppressed by some native Commissioners, whose policy can only be improved by the utmost publicity that can be given to it. The British South-African Colonist is still destined to play a large part in the fortunes of the Transvaal. He is the merchant of to-day, and will be the merchant of the future. He knows the wants and interests of the region, and understands both the Boer and the Kafir. He has not the prejudices of the Englishman fresh from home, and he appreciates and is better able to deal with the prejudices of those with whom he is thrown in contact. What we call middle-class social life is also largely supported by the colonial merchants, and they are the backbone of the nonconformist bodies in the Transvaal, of which the Wesleyans are far and away the strongest, and the Baptists making a beginning. Both in Natal and the Cape Colony, Boers were once numerous and are still found; so that the colonist has really never quite lost touch with the people who now only inhabit a country which resembles and adjoins his own. The youth of our two colonies are capable of supplying all the commercial clerks that may be needed for the commerce of the Transvaal; and if the Boers really understood their own interests, and were free from the domination of the Hollander, they would rely more on colonial friendship, and trouble themselves less with the idiosyncrasies of Downing Street. Common interests must produce fusion; imperialism perished with the disaster of Majuba; a confederation of South-African States already exists in men's hearts, it will soon reach their minds, and eventually be proclaimed by their mouths. Already, in the Transvaal, clear-headed men see that even commercial companies can neither be properly managed nor guided by a Board of Directors sitting in London; how little, then, can we expect to prevent the inevitable South-African Confederacy, which will be mainly composed of English-speaking people?

The Hebrew race is largely represented in the Transvaal, especially by those whose former home was in eastern Europe, and the Jew is destined to play a considerable and very influential part in the fortunes of the country. In a few years the Transvaal from being a purely geographical expression, inhabited by a pastoral community, has, by the utilization of its mineral wealth become a financial factor in the dealings of the Stock

Exchange at home, and an often regretted part in the
income-earning capital of private families. On this
bare South-African tableland fortunes have been made
by those who had nothing, and others have lost what
they had previously acquired elsewhere. Commercial
and mining companies were once of daily formation, as
though the whole country was one vast gold-reef, and
the Transvaal was to redress the financial balance of
Europe. The Jews have long possessed a genius for
dealing with precious stones and for being the best
financiers in the world. Diamonds brought them to
Kimberley, the discovery of gold-bearing reefs proved
at once a magnetic attraction to the Transvaal; they
largely created Johannesburg and its stock exchange—
now so silent,—and their element has proved a considerably constructive one in the formation of a commercial
community, many branches of which are now almost
entirely their own. With the untiring energy and
industry of the race, they have explored the whole country
in search of subjects for financial speculations, and their
knowledge of the Transvaal I estimate as far higher than
that of the Boers, who may, and doubtless do, excel
them in the possession of geographical details, but do not
approach their profound appreciation of the present and
future commercial capacity of the state. The Jew,
again, has a racial, but no particular political, nationality,
and thus can prosper with less suspicion and friction
amongst the burghers, who are naturally proud of the
development they see going on around them, yet know
it is not their work, and feel mistrust as to their future
independence in a purely Boer condition. And yet in
other respects the two races have little in common. No
one can deny that the Boer in his religion is a narrow
bigot, and not only in his heart dislikes unbelievers, but
would probably deny the right of a Jew or any pronounced heretic to hold an administrative part in the
Republic. On the other hand the Boer is a natural
sportsman, a pleasure which the Jew little appreciates,
who is at home in shop or counting-house, for which
the Boer has neither aptitude nor predilection. The

Jew also by his very cosmopolitanism becomes a good citizen, and some of the largest industries are being founded by him. His natural gaiety leavens the solemn national lump of Boer respectability. His literary abilities have largely contributed to the success of the Press, and in the Transvaal he is always "en évidence." I am speaking of the intelligent Jew, and not the scum of Houndsditch, which may also too plentifully be found, but which no more represents the race than numerous drunken ruffians who hail from Britain are to be taken as typical Englishmen. Like the travelling Christian, the migratory Jew does not let the rules of his creed sit very heavily on his shoulders; both eat at the same table of the same food, and there seems no particular restriction as to meat. Both creeds also afforded unique representatives. A Polish Jew who sat at my hotel table, and proved a very amusing companion, belonged to the most orthodox Hebrew sect; he was fairly learned in the Bible and Talmud, was of extreme and often violent orthodox-bigotry, but plainly admitted that his views had no claim on him in Pretoria, as Jews were only there to make money, and he certainly did not seek to remove that impression. I also knew a Hollander, who passed as a devout Christian, and who often told me that the Bible he read every day was the best of all books, and the New Testament his special delight. He also informed me of the different stages by which he was endeavouring to obtain a government appointment, in which the salary could be increased, not by bribes, but by what he more euphoniously called "additions that fell between the quay and the ship." But I regret to say that both of these acquaintances, the Jew and the Christian, had considerable doubts as to my orthodoxy, and regarded me with all the suspicion of "odium theologicum."

Commercial morality is a matter of constant evolution, subject to the stage of surrounding public opinion in which it exists. Some fortunes held in the Transvaal were mainly begun by the profit of buying diamonds from Kafirs who did not state the means by which they were obtained. Kafir robberies at the diamond-mines in time approached such large dimensions, that repressive penal acts were passed and enforced against these *indiscriminative* purchases. Consequently now on Cape Town breakwater may be seen convicts who arrived in the country too late for illicit diamond buying to be considered as one of the arts of a *clever speculator*. In a few years, even if it is not now the case, it will be considered bad taste to introduce the topic of amateur diamond purchases in some large, wealthy, and highly respectable South-African establishments. The "illicit diamond-buyer" is to-day the "company promoter," and public opinion, as soon as the law awakes, will equally approve of some professors of "flotation" joining their diamond-buying predecessors in undignified seclusion.

Our own countrymen form no inconsiderable portion of the Transvaal population; but the descendants of many will be of South-African birth, for there is an old and true proverb, "he who has once lived in South Africa will return to it again." When once the Transvaal is crossed by railways, the British farmer who is willing to permanently leave his old country and settle in what ought to be one of the finest farming regions of the world, will find a land worthy of his adoption. To the present time the resources of the Transvaal have only been sought beneath its surface, which remains practically untilled and untouched. The Boer farmer is simply a possessor of flocks and herds, and will probably remain so; the only hope of his being aroused from this deadly apathy, which keeps back the hands which register development on the clock of his country, is to encourage other farmers to settle in his midst, and show him what may be made of this wilderness. But the farmer must wait for the railway, and the railway will largely depend on the produce of the farmer. Johannesburg to-day is the most English town in the Transvaal; Pietersburg the most German; Pretoria the most cosmopolitan. One of the strangest features amongst the English is to find so many who would do equally well, if not better, at home. Many young fellows come out full of hope, who have had no other training but that most hopeless vocation of commercial clerk. Of course, some have succeeded in obtaining good positions, but others have almost patrolled the country,—sometimes a schoolmaster in a Boer's family, or the keeper of a small road-side store, seeking fortune as an inexperienced prospector, or even temporarily engaged as a waiter in an hotel; but you still hear no grumbling, but relief expressed that they have at least escaped the restrictions on life at home, breathe fresh air, and have less worry.

It certainly is a fact that no one seems to starve in the Transvaal; and it is equally true that men whose circumstances after a long stay in the country have become hopeless, if not desperate, still describe it as the finest and most improving land on earth. Whether it is that the knowledge of increased age and long absence from home have made return impossible, owing to precariousness of the livelihood they might expect to find, or whether it is the more free and untrammeled life led in Boerland, and the easy way by which men still, by some means or another, subsist, however bad their pecuniary resources may be, are questions that may perhaps be both answered in the affirmative. It is usual at some hotels to let the needy speculators and adventurers live on as boarders till better times arrive; and an acquaintance who once had a sleeping-share in an hotel told me that the consideration was not alway financially wrong. Whilst these indigent guests lived free, they advertised and recommended the hotel; the food was not missed when a large number of visitors had to be provided for, and in the changing fortunes of the country these derelicts frequently became once more able to pay their arrears. My experience was that every man obtained his subsistence by some means, though his affairs were in the blackest condition; even the "loafers" do not starve in South Africa. As I was told by an old trader who had traversed the country:—No one starves, for if such a thing did occur, it could only mean that the victim was either "a very bad lot" or of a very sensitive disposition. In the one case help might be withheld; in the other it is neither sought nor accepted.

It would be difficult for even the most fanatic opponent of Mr. Gladstone to imagine the deep, heartfelt, scornful detestation of his very name among many old British residents of this Republic. It is now but a decade since the Boer war—a war largely brought about by the arrogance and lost by the imbecility of many to whom British interests were then confided. I constantly met men who had risked life, fortune, and every hope in the cause of their old country, and who in the darkest day were prepared to go on, who, badly led, repined not, badly defeated were yet not beaten, but who have never forgiven what they call the infamy of the Gladstone surrender. I have found old English settlers, who took part and were ruined in the war, reviling the very name of their country; others who professing detestation of the Boers would yet help them —so they say—to fight against any renewed attempt at British supremacy; and all this not partial, not isolated, but common talk, which every traveller may hear who cares to mix with the people and listen to their views. I found it useless to argue; I had not the facts for defence. I recalled the old sugar-planting days of Malacca more than twenty years before, when my Scotch friends who managed the estates, and who were as a rule Tory and Jacobite to the bone, would angrily tell me they would travel twenty miles to see John Bright hanged.

The Kafir represents the labouring class of the Transvaal. Wherever manual unskilled work is required it is the Kafir who supplies it. He is the bricklayer's labourer, the porter, the miner, the farm hand, the shepherd, the scavenger, and even the common policeman. He promenades Pretoria in the most wonderful attire, for in the large towns he is not allowed to indulge in his primitive costume. His greatest glory is in the possession of an old soldier's tunic—numbers of which are imported from England for the Kafir trade; his most economic suit is a sack, through the bottom of which he makes three holes for the insertion of his head and arms. In the towns he is not allowed to

NATIVE POLICEMAN.

walk upon the paths, but must keep to the roads, and he is also required to retire from the streets and public thoroughfares when the Kafir bell is rung about 9 P.M. Our black brother may be despised, but the manual labour of the Transvaal at present depends upon him, and his labour is cheap and easily trained. His average wage is ten shillings per week and his "mealie" (ground maize); he is a cheerful worker, very apt to learn, and after my experience of Coolie labour in the East I have a great respect for the Kafir. He is fairly industrious, cheerful, frugal, and saving, for he has only travelled down to the white man's mart in order that he may in a few months have sufficient money to return and enjoy himself in the Kafir paradise. He does not understand the white man, who is always working to make money —and philosophically he is right; the European too often only looks upon him as a brute—when, philosophically, he is wrong. The two are old relations, who parted ages ago from the ancestral progenitors and who now meet each other again with a different colour of the skin but the same bodily structure; with a differentiation of ideas and development of brain, but with common animal instincts; and the Kafir from his heart believes his white brother to be the rich relation.

The Kafir relies principally upon meal for his food, but only too gladly partakes of meat when he can obtain it. On these occasions his culinary arrangements are best unseen. The head of a slaughtered ox is a great treat, so also is that of a sheep, which is cooked whole with wool, horns, and eyes included. As labourers they are distinctly clannish, and all my men once struck work because I proposed engaging some extra hands who belonged to a tribe who did not practice circumcision, so strong a hold does this rite have upon their social intercourse. They spend little of the wage-money they earn in the towns where they labour, but carry most of it back to their kraals and locations, and it has recently been estimated by one well-informed burgher that at least half a million of gold coinage, in sovereigns and half-sovereigns, are annually taken away by them, and thus by implication a large sum is withdrawn from circulation. The greater part of this money, however, eventually passes into the hands of the exclusively Kafir traders, who reside in their principal neighbourhoods, and thus again returns to the commercial heart of the Transvaal. Most of the Kafirs who labour in Pretoria and Johannesburg come down from the

Zoutpansberg district, and though I believe they would never be induced to fight for the Boers, they would still make excellent soldiers if properly trained, armed, and led by white officers. They could thus be led to storm a redoubt, though they would never prove material for forming Waterloo squares as targets for French cannon. The future development of the Transvaal depends greatly upon the Kafir, for in him centres the " labour question." Many undertakings already achieved would have been impossible had it not been for the cheapness with which unskilled labour could be obtained, and all calculations for large buildings or railway cuttings are greatly dependent on this factor, for the natural features of the wild country he has known so long will eventually be transformed by his hands.

The drainage and sanitary arrangements of Pretoria, when I visited it, would have been a disgrace to a country village at home, but great difficulties have been overcome and improvements are being made. It was somewhat singular to be strongly advised by all medical men not to drink any water, even if mixed with alcohol, without the same had been first strongly boiled and then filtered, for the water supplied to the town in open sluits was the most dangerous disseminator of typhoid malaria, the real danger of Pretoria and Johannesburg. No Good Templar ever more longed for a glass of pure water than I frequently did myself in a land where the smallest bottle of English ale is charged two shillings, and this, again, the doctors advised us not to drink, or at least seldom. The Boer has long since become one of the greatest coffee-drinkers on the face of the earth, and has found there is nothing better to be taken on his long wagon journeys; for those who use stimulants the coolest and mildest drink is a mixture of Cape hock and seltzer water; for the spirit-drinker, whiskey diluted with water. A company, however, has now supplied Pretoria with pure water, and drinking-fountains are erected in the town. There are three dangers to health in the Transvaal which may easily be avoided—want of cleanliness, intemperance, and the neglect of wearing warmer clothing after sunset, for weak lungs are liable to suffer from the sudden change of temperature that then ensues. But, on the other hand, the possessors of weak chests and moderately diseased lungs, to whom a European winter is a positive danger, may with a modicum of care live in physical ease and comfort breathing the glorious air of this high Transvaalian tableland. It was to the want of proper sanitary arrangements that the epidemic of 1889–1890 was doubtless due. This took the form of typhoid fever, with a frequent complication of pneumonia which attacked both lungs at the same time. Two judges, two doctors, and a large number of Europeans were quickly carried off, and the mortality was greater at Johannesburg than in Pretoria. At present the vital statistics are anything but satisfactory; the following figures, taken from a compilation made by Dr. Stroud, and published in the 'Press' newspaper, refer to the precincts of Pretoria alone:—

	1882 to 1887 inclusive.	1888.	1889.	1890.
Total deaths ..	162	123	169	171

Of the 171 deaths which occurred during 1890, 71 were men, 19 were women, and 81 were under 20 years of age.

Of the men	23	died between the ages of	20 and 30
,,	23	,, ,,	30 ,, 40
Of the women	6	,, ,,	20 ,, 30
,,	7	,, ,,	30 ,, 40

Of the 81 who died under 20 years of age, 7 died between the ages of 2 and 20. The remaining 74 died before the age of 2 years.

What the average of life amongst the Boer farmers may be I have no means of ascertaining, but I never saw many very aged individuals.

In Pretoria, as in Johannesburg, one met the queerest of social characters, and they comprised the army of company promoters, prospectors, financial agents, mining experts, and members of the various professions which enable money to be made on the credit of auriferous quartz being found in sufficient quantity to enable the formation of a limited liability company. But all these good people had one story to tell—they had all lost their money. Ask whom you would, converse anywhere with high and low alike, the knowing and unwary, the sharp and the dupe, all had lost and suffered at the collapse of the great bubble. The tale of the Johannesburg boom was all one heard; it was as though a mighty financial storm had raged and the shore was strewn with the bodies of these unhappy and disappointed speculators. It was a golden period, this era of the boom; anything served to float a company, and the price of shares rose daily. On all sides men purchased scrip, or held the shares they obtained by the process of flotation. At length a fall took place, owners still held, thinking the check temporary and that recovery must take place; but the collapse became sudden and severe and nothing could be saved of the fortunes so rapidly made—they existed on paper, and as paper they now remain. Cautious workmen who had saved a few hundred pounds were drawn into the vortex and lost their little all. A young baker once travelled with me in a coach who had managed to acquire by his business in the Transvaal some £3000, with which he returned to England. The echo of the boom brought him once more to the Transvaal with his money, every penny of which he lost, and he was when I met him working as a journeyman baker once more. The most extraordinary story of the vicissitudes of fortune in the Transvaal that I ever heard, related to two worthies who possessed together about £200 in cash and a wagon and oxen. They arrived at a canteen—many of which are a curse to the country—and absolutely drank and gambled away the whole of their money; they then sold the wagon and oxen for the same purpose and with a like result. They were soon penniless, and appealing to the keeper of the canteen where they had ruined themselves, induced that individual to help them, which he did with the donation of a sovereign, a bottle of spirit, and a loaf of bread, with which they started on an aimless tramp. The first day they swallowed the contents of the bottle to prevent—as described to me—the trouble of carrying it, the sovereign was soon spent, and their condition rapidly became desperate. At this moment they actually walked over an undiscovered gold-bearing reef near one of the mining centres of the country. They repaired to that town with samples of the quartz, borrowed sufficient of a financial agent to purchase the claim and secure the ground; a company was formed, and they returned home with £30,000 between them. I believe this story to be absolutely authentic.

In a small community like Pretoria the competition for social distinction is naturally very observable, and many seem to travel some thousands of miles from home to plant the arrows of outrageous respectability in African soil. Amongst other peculiarities is the adoption of our high-crowned white hat, which is only worn by doctors and lawyers, and is almost considered the exclusive privilege of those two professions. Thus a hat that was once considered to denote a somewhat sporting character in England, now marks the climax of legal and medical respectability in Pretoria. The President also considers it incumbent to wear a high-crowned black hat, but the other inhabitants have practically discarded the abomination.

I was much surprised to find so few well-kept gardens in a town which possessed so many Dutch inhabitants, and who might have been expected to have brought their flowering bulbs and love of gardening from their old country. I looked in vain for hyacinths or tulips, I scarcely ever saw a crocus, and never observed cultivated narcissi. These plants may possibly be found in the grounds of the richer inhabitants, but I certainly never found them in the gardens belonging to the houses of the more middle-class Dutch residents. The Hollander does not always carry his love of gardening to South Africa. The Briton is more apt to copy the old English home-scenes in the land of his adoption, and flowers spring up around his African home; his presence in the Transvaal is also denoted by the cricket-ground, and in the large towns by the race-course and grand stand, which seem to be inseparable signs of an English settlement in every part of the world I have visited. As time passes, and mutual antagonism and misunderstanding between Boer and Briton become more and more reduced, the British element in the Transvaal will be very considerable, and its descendants will form true colonials with little wish to return to their native land. Such men become good citizens, especially when they meet with a prosperity which was denied to them at home.

I once met a thriving Scotchman in Natal who gave me the history of his career in the colony. Twenty-seven years previously he had left a northern Scottish town, where he supported a wife and two children on a weekly wage of nineteen shillings. As he related the tale, "there were forty of us where I was employed and I was the best of the lot, but nineteen shillings a week was the most I could get." After his regular labour he worked at another occupation from 7 to 11 in the evening, and thus increased his small income. But at last he struck, he felt it was neither just nor honest he should always work like a slave in the mere effort of sustaining life, and he came to Natal as a government emigrant. The first day of his arrival he looked about, the second he obtained work and earned a sovereign, which he took home and justifying himself said: "There, wife, in one day in this country as much as I could earn in a week in Scotland!" Now he is a wealthy and prosperous man, and has been home for a trip to see the old land, where he cares not to live though he could now afford to do so. "No," as he remarked to me, "this is my country and my home; Natal has been *so kind* to me, and Scotland so different and *so hard.*" This is the true spirit of the colonist—there is a gratitude and love for his adopted land, and a stern resolve to protect her interests even if jeopardized by the mismanagement of the country that gave him birth. Each colony might have two representatives in the British parliament, to watch and explain purely colonial interests, whilst having the full privileges of ordinary members, and this could be made to in no way interfere with the present system of colonial government.

Whether these sturdy and life-long British colonists can be obtained for the Transvaal, to till its waste lands, increase its population, and develop its industries, or whether the Republic is still to be only a pastoral community, dependent on its auriferous quartz for an influx of foreigners with an uncertainty of speculative revenue, is for the Boer and Hollander to decide.

APPENDIX.

APPENDIX.

ENUMERATION AND DESCRIPTION

OF THE

ANTHROPOLOGICAL AND ZOOLOGICAL OBJECTS

COLLECTED BY THE AUTHOR

WITH CONTRIBUTIONS BY

Ernest E. Austen, Zool. Dept. Brit. Mus.
G. A. Boulenger, Zool. Dept. Brit. Mus., F.Z.S.
Jules Bourgeois, M.E.Soc.Fr.
J. H. Durrant, F.E.S.
C. J. Gahan, M.A., Zool. Dept. Brit. Mus.

Rev. H. S. Gorham, F.Z.S.
Martin Jacoby, F.E.S.
R. I. Pocock, Zool. Dept. Brit. Mus.
H. de Saussure, Socius hon. Soc. ent. Lond. Rossic. Belg. &c. &c.
Prof. C. Stewart, Pres.Linn.Soc. &c. &c.

And the Author.

The following pages are devoted to an enumeration of the zoological objects collected by myself in the Transvaal, and I have not added the names of any species described or recorded by other naturalists or travellers as from that country. The intention has been to avoid any approach to compilation, which can only be useful when much more material has been acquired and greater work done in different parts of the Transvaal. The fauna is distinctly South African, and has but very few elements of that belonging to the Eastern Tropical Region, and fewer still of that appertaining to the Western Tropics—a proof of which may be seen in the proportion of new species discovered, nearly all the others having previously been described from specimens collected much further south in the continent.

When the locality Pretoria is used as a habitat for a species, the town itself is not necessarily designated, but the district in which the town is situated, and this bears a similar relation to what the county of Yorkshire does to the city of York. I have also included the species procured in Natal, but this locality is always distinguished by *italics*. It must also be distinctly understood that the localities given are not intended to convey the impression that the species are not found elsewhere in the Transvaal, but only that at those spots I made my captures.

Many of the species have a very wide distribution, and I was surprised to meet with some old English friends, of which a list is added, though doubtless it could be considerably enlarged by further experience.

Species found in England and also by the Author in the Transvaal.

MAMMALIA.

Mus rattus.	Common Rat.

AVES.

Circus pygargus.	Montagu's Harrier.
Merops apiaster.	European Bee-eater.

COLEOPTERA.

Philonthus varians.	Black Staphylinid.
Dermestes vulpinus.	Destructive Beetle to hides and furs.
Aphodius lividus.	Dung-Beetle.
Corynetes rufipes.	Beetle frequenting bones and horns.
Corynetes ruficollis.	Beetle frequenting bones and horns.
Exochomus nigromaculatus.	" Ladybird " Beetle.

APPENDIX.

LEPIDOPTERA.

Pyrameis cardui.	"Painted-Lady Butterfly."
Acherontia atropos.	"Death's-head Moth."
Protoparce convolvuli.	"Convolvulus Hawk-Moth."
Deiopeia pulchella.	"Crimson-speckled Moth."
Sterrha sacraria.	"Vestal Moth,"
Nomophila noctuella.	"Pyral Moth."
Pyralis farinalis.	"Pyral Moth."

ORTHOPTERA.

Labidura riparia.	Earwig.
Phyllodromia germanica.	Cockroach.
Ectobia ericetorum.	Cockroach.

MAMMALIA.

MAMMALIA.

The vast herds of ruminants that once gave the Mammalian fauna of the Transvaal such a distinctive feature have now passed away, owing to ruthless and unrelenting destruction on the part of man, and the Carnivora will soon share the same fate. The lion is almost—if not entirely—confined to Zoutpansberg, and is becoming scarcer every year, while a good leopard-skin is much more difficult to obtain than was the case a few years ago. I procured the perfect skin and skull of a very fine young male lion, which was purchased in the Pretoria Market, but as its exact locality is doubtful, I have not included it in my list; and the same silence has been maintained as to several other skins purchased, but which may have belonged to animals killed by hunting-parties beyond the confines of the Republic. I paid most attention to the smaller mammals found near the town.

Excluding the valuable contribution of Prof. Stewart, the specimens have all been determined by Mr. Oldfield Thomas, and most of them I have placed in the collection of the British Museum.

PRIMATES.

ON SIX CRANIA, PROBABLY BELONGING TO THE MAKAPAN TRIBE, WATERBERG DISTRICT, TRANSVAAL, SOUTH AFRICA.

BY

C. STEWART,

Hunterian Professor of Comparative Anatomy and Physiology, and Conservator of the Museum of the Royal College of Surgeons; President of the Linnean Society, &c.

These six crania were obtained by Mr. W. L. Distant from the Makapan's Cave, in the Waterberg District, Transvaal, and have been kindly presented by him to the Royal College of Surgeons.

Three of the crania are those of adults, and three from young persons, the probable ages of the latter being about 12, 13, and 14 years.

All the crania are of the true dolichocephalic type, are platyrhine, except the one numbered 1299 E, and mesognathous except 1299 F.

The following gives the measurements of the crania and chief peculiarities.

1299 A. Metopic (persistent mid-frontal suture), showing the outward bulging of the frontal region of the temporal fossa commonly present in this condition, also a shallow depression immediately posterior and parallel to the coronal suture.

C 515. L 184. B 129. Bi 701. H 133. Hi 723. BN 105. BA 105. Ai 1000. Nh 47. Nw 25. Ni 532. Ow 37. Oh 34. Oi 919. Ca 1445.

1299 B. On both sides a deep fossa immediately below the inferior orbital foramen. Orbital height unusually great, viz. 38 mm.

C 510. L 185. B 128. Bi 692. H 128. Hi 692. BN 98. BA 100. Ai 1020. Nh 52. Nw 28. Ni 538. Ow 36. Oh 38. Oi 1056. Ca 1350.

1299 C.
C 504. L 182. B 121. Bi 665. H 129. Hi 179. BN 96. BA 95. Ai 990. Nh 43. Nw 26. Ni 605. Ow 37. Oh 32. Oi 865. Ca 1275.

1299 D. Aged about 14 years.
C 492. L 178. B 128. Bi 719. H 128. Hi 719. BN 95. BA 94. Ai 989. Nh 38. Nw 22. Ni 579. Ow 38. Oh 30. Oi 789. Ca 1305.

1299 E. Aged about 13 years. Mesorhine. Distance between outer borders of orbits small, viz. 85 mm.

C 500. L 181. B 131. Bi 724. H 128. Hi 707. BN 90. BA 91. Ai 1011. Nh 45. Nw 22. Ni 489. Ow 34. Oh 31. Oi 912. Ca 1530.

1299 F. Aged about 12 years. Orthognathous.
C 485. L 175. B 127. Bi 726. H 126. Hi 720.
BN 92. BA 87. Ai 946. Nh 36. Nw 22. Ni 611.
Ow 36. Oh 30. Oi 833. Ca 1340.

Average indices of the three adult crania :—

Breadth index (Bi) 686.
Height ,, (Hi) 708.
Alveolar ,, (Ai) 1003.
Nasal ,, (Ni) 558.
Orbital ,, (Oi) 948.

CARNIVORA.

Cynictis penicillata, Cuv.	Meer-Kat.	Pretoria.

INSECTIVORA.

Crocidura martensii, Dobs.	Large Shrew.	Pretoria.
Crocidura pilosa, Dobs.	Smaller Shrew.	Pretoria.

RODENTIA.

Myoxus murinus, Desm.	Dormouse.	Pretoria.
Mus rattus, Linn.	Rat.	Pretoria.
Mus coucha, A. Smith.	Mouse.	Pretoria.
Mus (Isomys) pumilio, Sparrm.	Striped Mouse.	Pretoria.
Pedetes capensis, Linn.	"Spring Haas."	Pretoria.

ARTIODACTYLA.

Pelea capreolus, Thunb.	Vaal Rehbok.	Pretoria.
Nanotragus scoparius, Schr.	"Oribi."	Pretoria.
Cephalolophus grimmii, Linn.	Duyker.	Spelonken, Zoutpansberg.
Cervicapra arundinum, Bodd.	Rietbok.	Pretoria.

AVES.

AVES.

I DID not succeed in finding an undescribed bird in the district of Pretoria, nor did I much expect to do so. The bare plains of the high veld support no rich avifauna, whilst the neighbouring districts of Lydenburg, Potchefstroom, the road to the Limpopo and the banks of that river had already been worked by those excellent field-ornithologists, Mr. Thomas Ayres * and Mr. F. A. Barratt †. Moreover, the high lands of Natal around Newcastle, which form part of the same area as that which comprises the Southern Transvaal, have been visited by Majors E. A. Butler and H. W. Feilden and Capt. S. G. Reid ‡.

Immediately around Pretoria the Accipitres are always *en évidence*. The Common Vulture, *Gyps kolbii*, as scavenger, continuously patrols the air, and settles in flocks as the carcass of some dead ox is sighted (see *ante*, pp. 69, 70). The Rufous Buzzard (*Buteo desertorum*) is a terror to the poultry-breeders around the town, and next to the Vulture is the most abundant in this order of birds. Montagu's Harrier (*Circus pygargus*) is not at all uncommon, but does not venture, as a rule, within a few miles of the town, and is difficult of approach. The Jackal Buzzard (*Buteo jakal*) is also very scarce in the district; I only procured it myself among the wooded lowlands of Zoutpansberg. My greatest acquisition was a specimen of Wahlberg's Eagle (*Aquila wahlbergi*), obtained a very few miles outside the town of Pretoria, a spot where the Black-shouldered Kite (*Elanus cœruleus*) could be generally seen flying or hovering high in the air, and seldom in reach of the gun. Several species of Kestrels were very abundant, usually fre-

---

* 'Ibis,' 1869, 1871, 1873, 1876-80.
† 'Ibis,' 1876.
‡ 'Zoologist,' 1882.

quenting rocky ground. The favourite food of *Cerchneis tinnunculoides* appears to be orthopterous insects, and Orthoptera and Coleoptera were found in the stomach of *Falco ruficollis*. For some facts relating to the Secretary-bird (*Serpentarius secretarius*) see *ante*, p. 68.

Some species very common in the wooded districts of Waterberg and Zoutpansberg are occasionally seen in the district of Pretoria, such as the Grey Plantain-eater (*Schizorhis concolor*) and the Yellow-billed Hornbill (*Lophoceros leucomelas*), examples of both of which were observed and obtained. Another bird not at all rare around Pretoria is the Golden Cuckoo (*Chrysococcyx cupreus*) ; in the stomach of one I found small Coleoptera, in that of another specimen small Orthoptera. Peters' Glossy Starling (*Lamprocolius sycobius*) and the Cape Glossy Starling (*Amydrus morio*) are very abundant in wooded rocky spots, and give a colour to the scene; while after the rains the Common Spreos (*Spreo bicolor*) assemble in flocks upon the veld, and devour the small Orthoptera there existing in great plenty.

Wherever wet places and high reeds are found, the Long-tailed Widow-bird (*Chera progne*) may usually be observed pursuing its laborious and difficult flight, heavily handicapped by its seasonally-developed tail, and is a good instance where sexual selection is exercised at the expense of protection.

Among the tamest of birds may be mentioned the Cape Long-claw (*Macronyx capensis*), which can frequently be killed when driving by a slash of the whip wielded by an expert Kafir, as a specimen in my collection was thus obtained. But this bird is not usually found around the outskirts of the town, as is that most friendly of visitors, the Cape Wagtail (*Motacilla capensis*), many of which fall a prey to small Dutch boys armed with that hideous instrument, the "catapult."

I give a list * of my captures, which may be taken to give a fair, but not exhaustive, estimate of the birds to be obtained around the capital of the Transvaal; and in preparing the same I must express my warmest thanks to Dr. R. Bowdler Sharpe,

* I have arranged this list according to the method pursued in Layard and Sharpe's 'Birds of South Africa.'

AVES. 165

who examined and identified the Accipitres, and to Capt. G. E. Shelley, who kindly went through and named the rest of the collection, excepting a few species identified by Mr. H. E. Dresser.

Order ACCIPITRES.

Gyps kolbii, Daud.	S.-African Griffon Vulture.	Pretoria.
Serpentarius secretarius, Scop.	Secretary-bird.	Pretoria.
Circus pygargus, Linn.	Montagu's Harrier.	Pretoria.
Astur polyzonoides, Smith.	Many-banded Goshawk.	Pretoria.
Buteo jakal, Daud.	Jackal Buzzard.	Spelonken, Zoutpansberg.
Buteo desertorum, Daud.	Rufous Buzzard (3 vars.).	Pretoria.
Aquila wahlbergi, Sundev.	Wahlberg's Eagle.	Pretoria.
Milvus ægyptius, Gm.	Yellow-billed Kite.	Pretoria.
Elanus cæruleus, Desf.	Black-shouldered Kite.	Pretoria.
Pernis apivorus, Linn.	European Pern.	Pretoria.
Falco ruficollis, Swains.	African Rufous-necked Falcon.	Pretoria.
Cerchneis rupicola, Daud.	South-African Kestrel.	Pretoria.
Cerchneis rupicoloides, Smith.	Large African Kestrel.	Pretoria.
Cerchneis tinnunculoides, Temm.	Lesser Kestrel.	Pretoria.
Cerchneis amurensis, Radde.	Eastern Red-footed Kestrel.	Pretoria.
Glaucidium perlatum, Vieill.	African Pearl-spotted Owlet.	Pretoria.
Asio capensis, Smith.	African Short-eared Owl.	Pretoria.
Strix flammea, Linn.	Barn-Owl.	Pretoria.

Order PICARIÆ.

Merops apiaster, Linn.	European Bee-eater.	Pretoria.
Melittophagus meridionalis, Sharpe.	Little Bee-eater.	Pretoria.
Coracias caudata, V.	Lilac-breasted Roller.	Spelonken, Zoutpansberg.
Lophoceros leucomelas (Licht.).	Yellow-billed Hornbill.	Pretoria.
Upupa africana, Bechst.	South-African Hoopoe.	Pretoria.
Schizorhis concolor, Smith.	Grey Plantain-eater.	Pretoria, Zoutpansberg.
Chrysococcyx cupreus (Bodd.).	Golden Cuckoo.	Pretoria.
Trachyphonus cafer, Less.	Levaillant's Barbet.	Pretoria.
Dendropicus cardinalis (Gm.).	Cardinal Woodpecker.	Pretoria.

Order PASSERES.

Turdus litsitsirupa, Smith.	South-African Thrush.	Spelonken, Zoutpansberg.
Pycnonotus layardi, Gurney.	Bulbul.	Pretoria.
Monticola brevipes, Strickl. & Scl.	Short-footed Rock-Thrush.	Pretoria.
Cossypha caffra (Linn.).	Cape Chat-Thrush.	Pretoria.
Myrmecocichla formicivora, Vieill.	Southern Ant-eating Wheatear.	Pretoria.
Saxicola monticola, Vieill.	Mountain Wheatear.	Pretoria.
Saxicola familiaris, Steph.	Familiar Chat.	Pretoria.
Saxicola pileata (Gm.).	Capped Wheatear.	Pretoria.
Pratincola torquata (Linn.).	S.-African Stone-Chat.	Pretoria.

AVES.

Nectarinia famosa, Linn.	Malachite Sun-bird.	Pretoria.
Zosterops virens, Bp.	Green White-eye.	Pretoria.
Parus afer, Gm.	South-African Titmouse.	Pretoria.
Hirundo semirufa, Sundev.	Red-breasted Swallow.	Pretoria.
Hirundo cucullata, Bodd.	Larger Stripe-breasted Swallow.	Pretoria.
Lanius collaris, Linn.	Fiskal Shrike.	Pretoria.
Lanius collurio, Linn.	Red-backed Shrike.	Pretoria.
Urolestes melanoleucus, Jard. & Selby.	S.-African Long-tailed Shrike.	Pretoria.
Laniarius gutturalis (P. L. S. Müll.).	Bacbakiri Bush-Shrike.	Pretoria.
Laniarius atrococcineus (Burch.).	Crimson-breasted Bush-Shrike.	Pretoria.
Telephonus senegalus (Linn.).	Common Red-winged Bush-Shrike.	Pretoria.
Nilaus brubru (Lath.).	Brubru Bush-Shrike.	Pretoria.
Bradyornis silens (Shaw).	Pied Wood-Shrike.	Pretoria.
Heterocorax capensis (Licht.).	African Rook.	Pretoria.
Corvus scapulatus, Daud.	White-bellied Crow.	Pretoria.
Lamprocolius sycobius, Hartl.	Peters' Glossy Starling.	Pretoria.
Spreo bicolor (Gm.).	Common Spreo.	Pretoria.
Amydrus morio (Linn.).	Cape Glossy Starling.	Pretoria.
Sitagra caffra (Licht.).	Weaver-bird.	Pretoria.
Hyphantornis velatus, Vieill.	Black-fronted Weaver-bird.	Pretoria.
Plocepasser mahali, Smith.	White-browed Weaver-bird.	Pretoria.

Vidua principalis (Linn.).	Common Widow-bird.	Pretoria.
Vidua ardens (Bodd.).	Red-collared Widow-bird.	Pretoria.
Chera progne (Bodd.).	Long-tailed Widow-bird.	Pretoria.
Pyromelana oryx (Linn.).	Red Bishop-bird.	Pretoria.
Estrelda astrild (Linn.).	Common Waxbill.	Pretoria.
Passer arcuatus, Gm.	Cape Sparrow.	Pretoria.
Poliospiza tristriata, Rüpp.	Three-streaked Grosbeak.	Pretoria.
Alæmon nivosa, Swains.	Cape Lark.	Pretoria.
Tephrocorys cinerea (Gm.).	Rufous-capped Lark.	Pretoria.
Macronyx capensis (Linn.).	Cape Long-claw.	Pretoria.
Motacilla capensis, Linn.	Cape Wagtail.	Pretoria.

Order COLUMBÆ.

Columba phæonota, Gray.	South-African Speckled Pigeon.	Pretoria.
Turtur senegalensis (Linn.).	Senegal Turtle-Dove.	Pretoria.
Pterocles gutturalis, Smith.	Yellow-throated Sand-Grouse.	Pretoria.

Order GALLINÆ.

Pternistes swainsoni (Smith).	Swainson's Francolin.	Pretoria.
Francolinus levaillantii, Temm.	Levaillant's Francolin.	Pretoria.
Francolinus gariepensis, Smith.	Orange River Francolin.	Spelonken, Zoutpansberg.
Francolinus subtorquatus, Smith.	Coqui Francolin.	Pretoria and Spelonken, Zoutpansberg.

AVES.

Order GERANOMORPHÆ.

Anthropoides paradisea (Licht.).	Blue Crane.	Pretoria.
Otis kori, Burch.	Kori Bustard or "Gom Paauw."	Pretoria.
Otis cærulescens, Vieill.	Blue Bustard.	Pretoria.
Otis afroides, Smith.	White-quilled Bustard.	Pretoria.

Order LIMICOLÆ.

Œdicnemus capensis, Licht.	South-African Thick-knee.	Pretoria and Zoutpansberg.
Glareola melanoptera, Nordm.	Black-winged Pratincole.	Pretoria.
Cursorius senegalensis, Licht.	Senegal Courser.	Pretoria.
Cursorius chalcopterus, Temm.	Bronze-winged Courser.	Pretoria.
Ægialitis tricollaris (Vieill.).	Treble-collared Sand-Plover.	Pretoria.
Ægialitis asiaticus (Pall.).	Asiatic Dotterel.	Pretoria.
Chettusia coronata (Temm.).	Crowned Lapwing.	Pretoria.
Machetes pugnax (Linn.).	Ruff.	Pretoria.

Order HERODIONES.

Bubulcus ibis (Linn.).	Buff-backed Egret.	Pretoria.
Scopus umbretta, Gm.	"Hammerkop."	Pretoria.

Order ANSERES.

Plectropterus gambensis (Linn.).	Spur-winged Goose.	Pretoria.

Order PYGOPODES.

Podiceps capensis, Bp.	Little Grebe.	Pretoria.

REPTILIA

AND

BATRACHIA.

REPTILIA AND BATRACHIA.

APART from the Python, of which I only heard accounts, the largest reptile I met in the Transvaal was the Monitor (*Varanus niloticus*), which was not uncommon on the banks of the spruits (see *ante*, p. 87). Another very common Lizard was *Agama hispida*, of which I have found four or five under a single stone; it runs about the bare veld and is easily caught. The most interesting species is *Mabuia trivittata*, which inhabits holes in banks in company with Toads, and, as already described, basks in the sun at the entrance to its hole, with its legs arranged close by the side of its body, which is curled up like a Snake, which the Lizard then much resembles (*ante*, p. 87). The Puff-Adder (*Vipera arietans*) I only found twice, and, strange to say, the two specimens were met with, at an interval of a fortnight, on exactly the same spot in a pathway at the foot of a cliff; on each occasion I nearly trod upon the reptile, which was basking in the dust under the midday sun. In walking through the high grass of the warm lowlands of Zoutpansberg one is frequently warned to be careful of Snakes; but in these excursions I only saw one individual, which was about six feet long, and sought refuge amidst some large blocks of quartzite before I could obtain a shot at it.

As the warm rainy season advances the silence of the veld (wherever accumulations of water are found) is broken by the croakings of Batrachians, and the hoarse bellow of the huge *Rana adspersa* makes the night hideous to those who live in the vicinity of the haunts of this handsome frog.

I have placed my small collection among the treasures in the British Museum, and that excellent authority, Mr. G. A. Boulenger, has contributed the following enumeration and notes, and described a new species of Snake which I found near Pretoria.

REPTILIA AND BATRACHIA.

By G. A. BOULENGER, F.Z.S. &c.

REPTILIA.

LACERTILIA.

Agama hispida, L.   Pretoria.
In the adult females the ear-opening is a little larger than the eye-opening, and the head is blackish, all over or on the sides only. One of the males shows a vermilion stripe along each side of the belly, from axilla to groin, and scattered spots of the same colour on the sides of the body and above the shoulder. All the specimens have a yellowish vertebral stripe.

Agama atricollis, Smith.	Pretoria.
Zonurus cordylus, L.	Pretoria.
Varanus niloticus.	Pretoria.
Nucras tessellata, Smith.	Pretoria.

A single young specimen. Black above, with three white lines along the back, and two series of round white spots along the sides; sides of head with black and white vertical bars; tail coralline red.

Eremias lineo-ocellata, D. & B.	Pretoria.
Gerrhosaurus flavigularis, Wiegm.	Pretoria.
Mabuia trivittata, Cuv.	Pretoria.

32 or 34 scales round the middle of the body.

Mabuia striata, Ptrs.   Pretoria.
The single specimen has 40 scales round the middle of the body, and is therefore referable to Peters's *M. wahlbergii*, which I now regard as not separable from *M. striata*.

Chamæleon parvilobus, Blgr.   Pretoria.

OPHIDIA.

Glauconia distanti, sp. n.

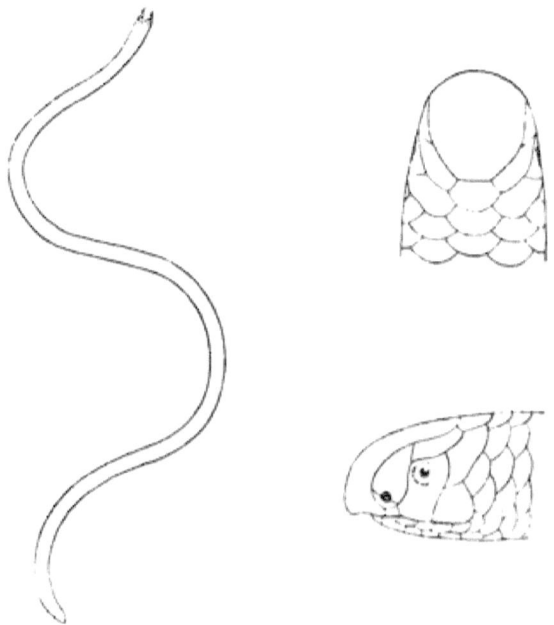

Glauconia distanti.

Snout rounded, projecting beyond the mouth, slightly hooked, the preoral portion concave inferiorly; supraocular present; rostral shield very large, extending posteriorly far beyond the level of the eyes, its upper portion nearly as broad as long, and covering almost the entire upper surface of the snout and crown; nasal completely divided into two, the lower part very small; ocular bordering the lip, between two labials, the anterior of which is very small; five lower labials. 14 scales round the body. Diameter of body 65 times in the total length, length of tail 12 times. Uniform blackish, the borders of the scales lighter.

A single specimen 130 millim. long. Pretoria.

In its hooked snout this species approaches *G. macrorhynchus*, Jan (Nubia), and *G. rostrata*, Bocage (Benguela and Angola). It differs from the former in its less slender form and larger rostral shield; from the latter in its rounded rostral edge and shorter tail.

Lamprophis rufulus, Licht.	Pretoria.
Leptodira rufescens, Gm.	Pretoria.
Psammophis sibilans, L.	Pretoria.
Causus rhombeatus, Licht.	Pretoria.
Vipera arietans, L.	Pretoria.

BATRACHIA.

Rana natalensis, Smith.	Pretoria.
Rana adspersa, Bibr.	Pretoria.

Bottle-green above, yellow beneath; base and inner surface of limbs orange.

Rana angolensis, Bocage.	Pretoria.
Phrynobatrachus natalensis, Smith.	Pretoria.
Bufo regularis, Reuss.	Pretoria.
Bufo carens, Smith.	Pretoria.

The lateral glandular fold and the larger glands on the sides of the body and on the limbs bright red. Tympanum much smaller in the young than in the adult.

ARACHNIDA

AND

MYRIOPODA.

ARACHNIDA AND MYRIOPODA.

All the species of Acari which I found—several of which remain unidentified—were taken during the dry season under stones. At this time few animals frequent the dried and parched veld, and it is probable that these parasites then hibernate.

Solpuga chelicornis, despite its formidable appearance, is attacked by small birds. I once witnessed the Solpuga trying to escape from the persistent attacks of a bird which appeared to be the Cape Wagtail (*Motacilla capensis*), and I eventually secured the pursued specimen.

The new species *Nephila transvaalica*, Pocock, is abundant during the rainy season in the wood-bush. It lives in small colonies in gigantic webs (*ante*, p. 91).

The large Myriopod, *Spirostreptus transvaalicus*, a new species now described by Mr. Pocock, is very rare. I obtained two specimens at Pretoria, and no one to whom I showed them had ever seen such a large species before.

ARACHNIDA.

By R. INNES POCOCK, Zool. Dept., Brit. Mus.

The following is a list of the Arachnida obtained by Mr. Distant in the Transvaal.

ACARI.

Of this order several examples of the family Ixodidæ were discovered under stones. To only one of these, however, namely *Amblyomma hebræum* of Koch, I am able to assign a name. The rest appear to be referable to the same genus *Amblyomma*.

SOLPUGÆ.

Solpuga chelicornis, Licht. & Herbst.   Pretoria.

ARANEÆ.

Fam. SPARASSIDÆ.

Ocypete megacephala, C. Koch.   Pretoria.

I provisionally only refer this species to C. Koch's old genus *Ocypete*. It does not appear to fall within the limits of any of the genera characterized by Mons. Simon in his paper on this family.

Fam. EPEIRIDÆ.

Gastracantha, sp.   Pretoria.

Argiope nigrovittata, Thorell.   Pretoria.

A widely distributed S.-African species.

Nephila transvaalica, sp. n.   (Tab. V. fig. 4.)   Pretoria.

Cephalothorax black, clothed with silver-white hairs; palpi flavous, with the apical segment fuscous; first two pairs of legs with coxæ, femora, and patellæ black; tibiæ adorned with four alternating bands, two flavous, two black, the proximal extremity being black, the distal flavous; tarso-metatarsus black, with only its proximal end narrowly flavo-annulate; third and fourth pairs of legs mostly black, the proximal three-fourths of the femur, however, flavous; maxillæ and labium black, the inner border of the maxilla and the apex of the labium flavous; mandibles black; sternum flavous, except for a very fine black marginal line; abdomen with a lateral band of silver hairs and marked above with five transverse flavous bands.

Cephalothorax furnished with two conspicuous tubercles; its lateral margins smooth.

Sternum cordate, strongly narrowed posteriorly, with a low tubercle at the base of the labium and at the base of each of the two anterior pairs of legs; also a minute tubercle on each side of the posterior extremity of the sternum.

Legs: femora of the first two pairs aciculate; tibiæ of these same legs with a tuft of hairs on the black band; tibia of the fourth pair covered with long hairs throughout; the tarso-metatarsus of the fourth pair hairy at the proximal end.

Measurement in millimetres.—Length of cephalothorax 9, width 6, width of cephalic portion 5; length of abdomen 12, width 8·5; length of anterior leg 46, of posterior leg 37.

Two dried female examples.

This species is closely related to *N. annulata* of Thorell; but of this last-named form I have only seen a very brief description. (R. I. P.)

MYRIOPODA.

By R. INNES POCOCK, Zool. Dept., Brit. Mus.

DIPLOPODA.

Fam. GLOMERIDÆ.

Sphærotherium obtusum, C. Koch.   Durban, Natal.

Fam. IULIDÆ.

Spirostreptus meinerti, Porath.   Pretoria.

Spirostreptus meinerti, Porath, Œfv.Vet.-Akad. Förhand. 1872, no. 5, p. 37.

This species was described originally from Caffraria. Mr. Distant obtained a single female specimen.

Spirostreptus transvaalicus, sp. n.   Pretoria.

♂. *Colour*: head piceous, labral region obscurely ferruginous; antennæ fuscous; legs deep ochraceous; somites polished black behind, pale anteriorly. *Head* smooth above, with a feeble vertical sulcus, strongly rugose with striæ and punctures below; with 6 or 7 punctures above the angular labral excision. *Antennæ* slender, longer than the head by the apical segment; the second segment the longest, the third, fourth, and fifth subequal in length. *Eyes* nearly twice as wide as long, narrowed internally, and separated by a space which is a little greater than a diameter, composed of about 80 ocelli arranged in about 13 vertically oblique series. The process on the mandible subacute.

The first tergite smooth above, the lateral portion marked with two complete and several short grooves; other grooves following the curvature of the anterior angle, the posterior angle of the lateral portion rounded and obtuse, the anterior angle strongly produced forwards into an apically rounded process, the lower margin of which is slightly convex, and the

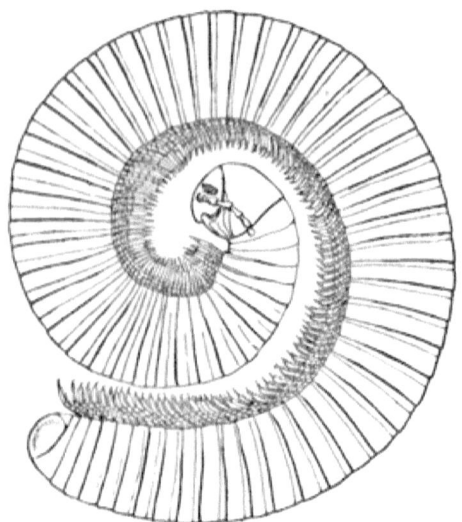

Spirostreptus transvaalicus.

upper, which is continuous with the anterior margin of the tergite, concave. The rest of the somites with their posterior portion entirely smooth and polished above, finely punctulate at the hinder end of the body, longitudinally striate below, the anterior portion finely striate transversely. Sternal areas nearly smooth; ventral grooves short. The pores, a little below the middle of the side, minute, and at the hinder end of the body almost invisible. Anal somite small; the tergite mesially angled above, scarcely covering the upper angle of the valves, marked at the base of the angle by a shallow punctulate constriction; valves convex, with their margins very lightly compressed and punctulate; sternite triangular.

Legs with the fourth and fifth segments padded beneath, the first and second hairy above at their distal extremities, the distal segment with a single large spine above the claw, smaller spines on each side of the claw and two irregular series of spines along the lower surface. The copulatory feet with the anterior piece narrowed below, marked by two deep grooves separated by a keel, terminating inferiorly in a smooth rounded prominence, the posterior piece terminating below in a prominence which is somewhat similar, but pointed below. The appendage consists of two pieces, a shorter, straight, simple pointed style, and a long curved flagellum, which is expanded mesially, pointed and simple at the end, but which gives off a short, simple, slender, curved process at the distal end of the expanded part of its length.

Number of somites 70.

Length about 200 millim. (about 8 English inches).

Two male specimens were obtained.

This handsome species is very closely allied to *Sp. heros* of Porath from Caffraria. The two, however, appear to differ slightly, at least in colour, and *Sp. heros*, according to the description, bears no secondary process on the flagellum of the appendage of the copulatory feet. (R. I. P.)

Spirostreptus (Odontopyge) pretoriæ, sp. n.   Pretoria.

Colour: head fuscous above, ferruginous beneath the line of the antennæ; antennæ with the first, second, third, and fourth segments ferruginous, the fifth fuscous distally, the sixth wholly fuscous; the first tergite deep slate-grey, with its anterior and posterior margins flavous; the rest of the somites with the posterior border widely flavous; the middle part of the somite slate-grey, laterally fading above and below into a paler ferruginous grey; anal somite wholly fuscous; legs ferruginous.

Head almost smooth above, finely punctulate, with a feeble sulcus, rugose and punctured on the labral region. Eyes widely separated, triangular, each consisting of about 50 ocelli. First tergite almost entirely smooth, the lateral portion subrectangular, with an anterior marginal sulcus and a second sulcus running from above the anterior angle to the posterior angle. The rest of the somites marked with a complete and strong transverse sulcus, the area behind this sulcus weakly longitudinally sulcate below; the whole of the upper surface very finely and closely rugulose, being marked with minute abbreviated, irregular, longitudinal striolæ. The pores conspicuous, in the middle of the hinder half of the somites, halfway up the side. Anal somite closely punctulate and reticulated throughout, the caudal process compressed; the valves convex, not compressed, but with a strong marginal sulcus. The teeth at the summit of the valves nearly erect, spiniform, almost in the same line as the posterior margin of the valves. The fine fringe along the hinder margin of the tergites perfectly even.

Number of somites 63.
Length about 45 millim. (*R. I. P.*)

INSECTA.

INSECTA.

COLEOPTERA.

GEODEPHAGA.

SOUTH AFRICA is well represented by this Tribe of Coleoptera, and the Transvaal is no exception to the rule. The genus *Manticora*, so restricted to this region, I only found in one species around Pretoria. It is very local, appears shortly after the first rains, and is very numerous indeed on the restricted areas it frequents, where it actively forages about the bare veld. It is, however, found only for a few weeks,—at least such was my experience. *Dromica* is a scarce genus, and the only species met with, *D. gigantea*, was seen in solitary examples at scattered intervals during the wet season. *Polyhirma macilenta* is very common on the paths and roadways of the open veld; but *Cypholoba ranzanii* is a much rarer species, and I only took it in Zoutpansberg. The two most common species of the genus *Anthia* are *A. thoracica* and *A. maxillosa*, but there are nothing like the number of species in the Transvaal as can be found in other parts of Southern Africa. *Atractonotus formicarius* is found under stones on the hardest and driest part of the veld, whilst under stones in damp places I found the rare *Callistomimus sexpustulatus*, *Crepidogaster bimaculatus*, and *Chlænius ritticollis*. But the streams in the summer are so violent, and the stones on the banks so frequently overturned and washed from their places, that it is very difficult to find the homes of these small Carabidæ.

For the identification of many of the species of Carabidæ I am indebted to that high authority Mr. H. W. Bates.

Fam. CICINDELIDÆ.

Manticora tuberculata, De Geer.	Pretoria.
Dromica gigantea, Brême.	Pretoria.

Fam. CARABIDÆ.

Acanthogenius dorsalis, Klug.	Pretoria.
Triænogenius corpulentus, Chaud.	Pretoria.
Pheropsophus litigiosus, Dej.	Pretoria.
Pheropsophus fastigiatus, Linn.	Pretoria.
Brachinus armiger, Dej.	Pretoria.
Crepidogaster bimaculatus, Boh.	Pretoria.
Calleida angustata, Dej.	Pretoria.
Hystrichopus caffer, Illig.	Pretoria.
Lebia, sp. ?	Pretoria.
Tetragonoderus bilunatus, Klug.	Pretoria.
Graphipterus cordiger, Dej.	Pretoria.
Graphipterus westwoodi, Brême.	Pretoria.
Graphipterus ovatus, Pering.	Pretoria.
Anthia thoracica, Fabr.	Pretoria.
Anthia maxillosa, Fabr.	Pretoria.
Anthia mellyi, Brême.	Pretoria.
Anthia æquilatera, Kl.	Pretoria.
Anthia desertorum, Thom.	Pretoria.
Anthia, sp. ?	Pretoria.
Piezia angusticollis, Boh.	Pretoria.
Cypholoba ranzanii, Bertol.	Zoutpansberg.
Polyhirma macilenta, Oliv.	Pretoria.
Atractonotus formicarius, Erichs.	Pretoria.
Scarites rugosus, Wied.	Pretoria.
Clivina grandis, Dej.	Pretoria.
Clivina, sp. ?	Pretoria.
Chlænius subsulcatus, Dej.	Pretoria.
Chlænius cylindricollis, Dej.	Pretoria.
Chlænius vitticollis, Boh.	Pretoria.
Callistomimus sexpustulatus, Boh.	Pretoria.
Harpalus capicola, Dej.	Pretoria.
Harpalus, sp. ?	Pretoria.
Hypolithus, sp. ?	Pretoria.
Abacetus obtusus, Boh.	Pretoria.
Euleptus caffer, Boh.	Pretoria.
Megalonychus interstitialis, Boh.	Pretoria.

Fam. GYRINIDÆ.

Aulonogyrus abdominalis, Reg. (see *ante*, p. 45). — Pretoria.

Fam. HYDROPHILIDÆ.

Hydrous, sp. ? — Pretoria.

Fam. STAPHYLINIDÆ.

Of the very few species of this family that were collected, the most interesting is *Philonthus varians*, a wide-ranging beetle found in England. For the identifications I am indebted to Dr. D. Sharp, who has special knowledge of the family.

Philonthus punctipennis, Woll.	Pretoria.
Philonthus varians, Payk.	Pretoria.
Philonthus, sp. ?	Pretoria.
Staphylinus hottentottus, Nordm.	Pretoria.

Fam. PAUSSIDÆ.

Pentaplatarthrus natalensis, Westw. — Pretoria.

Fam. SILPHIDÆ.

I have enumerated the following species under the names they bear in the collection of the British Museum. *Silpha pernix*, Burch., appears to be a MS. name; I have seen the type specimen, but am quite unable to trace the description.

Silpha pernix, Burchell, MS.?	Pretoria.
Silpha capensis, Dej.	Pretoria.

Fam. HISTERIDÆ.

I found very few members of this family around Pretoria, though I am informed that 213 species are already recorded from South Africa. Though the habits of the Histeridæ are stercoraceous, I found several species under stones on the bare veld during the dry season.

My friend Mr. George Lewis, who has made the Histeridæ his special study, has identified the specimens here enumerated.

Hister caffer, Erichs.	Pretoria.
Hister fossor, Erichs.	Pretoria.
Hister hottentotta, Erichs.	Pretoria.
Hister ovatula, Mars.	Pretoria.
Saprinus gabonensis, Mars.	Pretoria.
Saprinus natalensis, Fabr.	Pretoria.

Fam. DERMESTIDÆ.

The universally distributed *Dermestes vulpinus* was found in great abundance at Pretoria.

Dermestes vulpinus, Fabr.	Pretoria.
Æthriostoma gloriosa, Fabr.	Pretoria.
Attagenus, sp. ?	Zoutpansberg.

SCARABÆIDÆ.

Fam. COPRIDÆ.

The habits of the Copridæ are so uniform in character that little can be added to what has already been written on the subject. They are mostly strong fliers, perhaps in length of flight unequalled by any other Transvaalian Coleoptera, and this particularly applies to the genera *Scarabæus*, *Sebasteos*, and *Gymnopleurus*. The giant *Heliocopris* is scarce and fond of elevated situations. The commonest species is *Onilicellus militaris*, which may be found in almost every deposit of dung.

For the correct identification of several species I am much indebted to the assistance of Mr. C. O. Waterhouse, of the British Museum; and I should not have ventured to describe the new species of *Bolloceras*, but for the help of my friend Mr. H. W. Bates.

Scarabæus convexus, Hausm.	Pretoria.
Scarabæus savignyi, McLeay.	Pretoria.
Scarabæus nigræneus, Boh.	Pretoria.
Scarabæus hottentottus, McLeay.	Pretoria.
Scarabæus bohemani, Harold.	Pretoria.

Scarabæus interstitialis, Boh.	Pretoria.
Sebasteos galenus, Westw.	Pretoria.
Gymnopleurus wahlbergi, Fåhr.	Pretoria.
Gymnopleurus cælatus, Wied.	Pretoria.
Gymnopleurus, sp. ?	Pretoria.
Coptorhina klugii, Hope.	Pretoria.
Heliocopris hamadryas, Fabr.	Pretoria.
Catharsius, sp. ?	Pretoria.
Copris fidius, Oliv.	Pretoria.
Copris contractus, Boh.	Pretoria.
Onitis caffer, Boh.	Pretoria.
Onitis, sp. ?	Pretoria.
Phalops flavocinctus, Klug.	Pretoria.
Onthophagus gazella, Fabr.	Pretoria.
Onthophagus, spp. ?	Pretoria.
Oniticellus militaris, Castelu.	Pretoria.
Aphodius lividus, Oliv.	Pretoria.
Aphodius, spp. ?	Pretoria.
Hybosorus arator, Illig.	Pretoria.
Bolboceras batesii, sp. n.	Pretoria.
Trox, sp. ?	Pretoria.

Description.

Bolboceras batesii, sp. n. (Tab. I. fig. 5.)

Castaneous; antennæ ochraceous. Clypeus almost black, very thickly and coarsely punctate, with a short dentiform prominence or tubercle at the centre of the anterior margin, a similar tubercle on each lateral margin at its narrowing angle, and two tubercles side by side, on the middle of the forehead; posterior margin of the head narrowly ochraceous. Thorax moderately thickly punctate, with four obscure and obtuse bosses on its anterior slope, its posterior angles strongly rounded and prominent. Scutellum thickly and somewhat finely punctate, its margins narrowly blackish. Elytra longitudinally striate, the striations roughly punctate; median suture and extreme outer margin somewhat blackish. Legs beneath and margins of body strongly and ochraceously hirsute, with a prominent tuft at base of antennæ.

Long. 13 millim., lat. 9½.

Distinguished by the short dentiform prominence at centre of front margin of clypeus, and by the two tubercles to the forehead. (*W. L. D.*)

Fam. MELOLONTHIDÆ.

A considerable number of the flower-visiting species of this family are found in the Transvaal, but in nothing like the number in which they abound in more Southern Africa. Species of the genera *Eriesthis, Pachycnema, Monochelus,* and *Dichelus* are found on the few scattered flowers that bloom on the veld, and *Eriesthis semihirta* is certainly the most abundant and common species.

The smaller species of the family are still so little worked out that I can only make a small enumeration of those that have been described.

Eriesthis semihirta, Burm.	Pretoria.
Eriesthis guttata, Burm.	Pretoria.
Pachycnema tibialis, Oliv.	Pretoria.
Monochelus, sp. ?	Pretoria.
Dichelus vulpinus, Burm.	Pretoria.
Serica, spp. ?	Pretoria.
Trochalus, sp. ?	Waterberg.
Schizonycha, sp. ?	Pretoria.

Fam. RUTELIDÆ.

To the difficult species of this family the same remark applies as to the Melolonthidæ—they are little worked out. I found, as I did in Malacca, that many species come to light, and are almost only found in that way. The peculiar habits of two species of *Adoretus* have already been described (*ante,* p. 47).

Popillia bipunctata, Fabr.	Durban, Natal.
Adoretus luteipes, Casteln.	Pretoria.
Adoretus, sp. ?	Pretoria.

Fam. DYNASTIDÆ.

I did not meet with many of this family in the Transvaal. *Oryctes boas* was very abundant in old tan, and in its larval condition is, I have little doubt, eaten by the Wagtail (*Motacilla capensis*), as numbers of these birds were always searching the material which contained the larvæ.

Heteronychus, sp.?	Pretoria.
Oryctes boas, Fabr.	Pretoria.
Cyphonistes rallatus, Wied.	Pretoria.
Syrichthus verus, Burm.	Pretoria.

Fam. CETONIIDÆ.

I paid considerable attention to the collection of these beetles, for they appeared with the flowers, and as plants and trees successively blossomed, so new species of Cetoniidæ were found upon the bloom. Often the time of the appearance of the insect was as limited as the duration of the flower. *Anoplochilus tomentosus* is found on the open veld, buried in the bloom of a dicotyledonous plant somewhat resembling our own Dandelion. The blooms of Asclepiads (*Gomphocarpi*) are visited by *Melinesthes umbonata*, species of *Oxythyrea* (including the widely distributed *Oxythyrea cinctella*), *Coptomia umbrosa*, and *Tephræa morosa*. *Diplognatha silicea* is of very strong flight and I only secured it on the wing, and in the same manner I took the rare *Ischnostoma nasuta*. The two commonest species are *Plæsiorrhina plana* and *Pachnoda flaviventris*; both are found nearly throughout the whole of the summer, and their depredations on apples in the Natal Colony have already been described (*ante*, p. 126).

I have to thank Mr. Oliver Janson for assisting me in the identification of some species of this family.

Hypselogenia concava, Gory & Perch.	Pretoria.
Diceros algoensis, Westw.	Pretoria.
var. *flavipennis*, Westw.	Pretoria.
Ischnostoma nasuta, Schaum.	Pretoria.
Plæsiorrhina plana, Wied.	Pretoria.

Anisorhina flavomaculata, Fabr., var. *egregia*, Boh.	Pretoria.
Melinesthes umbonata, Gory & Perch.	Pretoria.
Coptomia umbrosa, Gory & Perch.	Pretoria and Durban, Natal.
Elaphinis irrorata, Fabr.	Durban, Natal.
Elaphinis latecostata, Boh.	Durban, Natal.
Gametis balteata, De Geer.	Pretoria.
Anoplochilus tomentosus, Gory & Perch.	Pretoria.
Oxythyrea amabilis, Schaum.	Pretoria.
Oxythyrea perroudi, Schaum.	Pretoria.
Oxythyrea cinctella, Schaum.	Pretoria.
Oxythyrea rubra, Gory & Perch.	Pretoria.
Oxythyrea hæmorrhoidalis, Fabr.	Pretoria.
Oxythyrea æneicollis, Schaum.	Pretoria.
Oxythyrea marginalis, Swartz.	Pretoria.
Tephræa morosa, Schaum.	Pretoria.
Trichostetha placida, Boh.	Durban, Natal.
Aplasta dichroa, Schaum.	Pretoria.
Cetonia cincta, De Geer.	Pretoria and Charlestown, Natal.
Pachnoda flaviventris, Gory & Perch.	Pretoria and *Natal*.
Pachnoda leucomelana, Gory & Perch.	Pretoria.
Rhabdotis aulica, Oliv.	Pretoria.
Rhabdotis sobrina, Gory & Perch.	Pretoria.
Rhabdotis semipunctata, Fabr.	Pretoria.
Diplognatha silicea, McLeay.	Pretoria.
Diplognatha hebræa, Oliv.	Pretoria.
Macroma cognata, Schaum.	Durban, Natal.
Ptychophorus leucostictus, Schaum.	Pretoria.
Spilophorus plagosus, Boh.	Pretoria.
Hoplostomus fuligineus, Oliv.	Pretoria.

Notes.

Diceros flavipennis, Westw., is certainly only a varietal form of *D. algoensis*, Westw. The species is very local, only found on the leaves of one particular shrub with an insignificant bloom, and both forms occur together.

Oxythyrea hæmorrhoidalis, Fabr.

A variety is very common around Pretoria in which the ground-colour of the elytra is shining bluish, and not green; a scarcer variety is found in which the colour is rosy red.

Rhabdotis aulica, Oliv.

I took a variety of this species in which the upper surface is rosy red, not green.

SERRICORNIA.

Both Buprestidæ and Elateridæ are very scarce at Pretoria, the bare plains being apparently uncongenial habitats for the group. Still the largest and handsomest species of the Transvaal Buprestidæ (*Sternocera orissa*, var.) is found on the few scattered Acacias which are thinly distributed on the veld, as before described (*ante*, p. 91).

Elateridæ are probably much more numerous, and the few species here enumerated give little idea of this Coleopterous family as found even on the high veld.

My best thanks are due to Dr. Caudèze, of Liège, for his determination of these obscure Elaters.

Fam. BUPRESTIDÆ.

Sternocera orissa, Buq., var. *liturata*, White.	Waterberg and Pretoria.
Amblysterna vittipennis, Boh.	Zoutspansberg.
Scaptelytra sulphureovittata, Fåhr.	Durban, Natal.
Psiloptera gregaria, Fåhr.	Pretoria.
Psiloptera viridimarginata, Fåhr.	Zoutpansberg.
Psiloptera calamitosa, Fåhr.	Zoutpansberg.
Chalcogenia cuprea, Lap. & Gory.	Pretoria.
Melybæus crassus, Lap. & Gory.	Pretoria.

Fam. ELATERIDÆ.

Aliteus adspersus, Herbst.	Pretoria.
Psephus puncticollis, Boh.	Pretoria.
Heteroderes inscriptus, Erich.	Pretoria.
Cardiophorus præmorsus, Boh.	Pretoria.
Pleonomus wahlbergi, Caud.	Pretoria.

MALACODERMATA.

Fam. LYCIDÆ.

These singular beetles are found in considerable profusion in the few small wooded spots to be met with around Pretoria. They are often seen in great quantity on the leaves of shrubby trees, and by their sluggish habits and bright coloration seem to fear few enemies, an idea accentuated by the fact that they are sometimes mimicked by other beetles, such as the Longicorn *Amphidesmus analis*.

I was able to secure twelve species of the genus *Lycus*, and in their identification have had much valuable assistance from Mons. Bourgeois, of Alsace, who has specially studied the Malacodermata, and has here described a new species.

Lycus dilatatus, Dej.	Pretoria.
Lycus subtrabeatus, Bourg.	Pretoria and *Durban, Natal*.
Lycus bremei, Guér.	*Durban, Natal*.
Lycus rostratus, Linn.	Pretoria.
Lycus æolus, Murray.	Pretoria.
Lycus constrictus, Boh.	Pretoria.
Lycus pyriformis, Murray.	Pretoria.
Lycus ampliatus, Fåhr.	Pretoria.
Lycus zonatus, Fåhr.	Pretoria, Zoutpansberg, and *Durban, Natal*.
Lycus integripennis, Bourg.	Pretoria.
Lycus distanti, sp. n.	Pretoria.
Lycus kolbei, Bourg.	Pretoria.

Description.

Lycus distanti, sp. n. (Tab. I. fig. 3.)

Lycus (*Chlamydolycus*) *distanti*, Bourg.

♂. A *L. elevato*, Guér.-Mén., cujus vicinus, rostro paullo breviore, elytrorum expansione latiore, magis rotundata, clytris ipsis apicem versus magis attenuatis costisque primis et secundis elevatis, bene distinctis præcipue differt.

Subcordatus, glaber, nitidiusculus, supra ochraceus, thoracis disco plaga triangulari apicem haud attingente, elytrorum regione scutellari trienteque apicali nigris; subtus niger, nitidior, trochanteribus femorumque stirpe sicut et abdomine

(medio excepto) ochraceis; prothorace leviter transverso,
subquadrato, antice late rotundato, lateribus fere parallelis,
disco longitudinaliter canaliculato, anticis posticis rectis,
haud productis; elytris ante medium in expansionem magnam
concolorem, supra concavatam et valde reflexam, infra autem
convexans et declivem rotundato-ampliatis, dein apicem
versus attenuatis ibique singulatim rotundatis, irregulariter
punctato-reticulatis, costis 2 elevatis, bene distinctis, apice
abbreviatis ornatis; abdominis segmento penultimo integro,
ultimo elongato-triangulari, bivalvato, forcipe apicem versus
attenuato ibique leviter curvato, simplici.

Long. 13 mill.; lat. thorac. 3 mill.; lat. max. elytr. 11 mill.

♀. *Hucusque invisa.*

Hanc speciem insignem Dom. W. L. Distant, qui eam detexit,
dicare gaudeo. (*J. Bourgeois.*)

For notes relating to the few beetles belonging to the two following families, Lampyridæ and Melyridæ, I am indebted to the Rev. H. S. Gorham.

Fam. LAMPYRIDÆ.

Luciola capensis, Oliv. Ent. ii. no. 28, p. 21 (*Lampyris*). Pretoria.

A *Luciola* with the thorax, metasternum, scutellum, coxæ, and femora yellow, the disk of the thorax with a large pitchy black mark narrowly divided by the yellow carina and a central spot. The elytra are leaden black with the suture brownish. I have seen this insect named as above, but I doubt if it is the *L. capensis* of Fabricius or of Olivier. A single male.

(*H. S. Gorham.*)

Fam. MELYRIDÆ.

Hedybius amœnus, sp. n. (Tab. I. fig. 2.)

Læte flavus, capitis basi, thoracis macula parva, antice excisa,
abdominis apice tarsisque posticis nigris; antennarum articulis sex ultimis nigro-notatis,' elytris et metasterno cærulcis.
Long. 5 millim. Pretoria.

Two specimens. A species to be distinguished in its genus by the single black thoracic spot, which, however, looks rather like two oblong spots united, and by the colour of its legs and antennæ, among other points. (*H. S. Gorham.*)

Anthocomus ? sp. ?   Pretoria.

One example of a very small Melyrid beetle pertains, I think, to this genus.

Fam. CLERIDÆ.

In this family two species of *Corynetes* are recorded, both well known in Britain; they were found among accumulations of horns.

Corynetes rufipes, Fabr.	Pretoria.
Corynetes ruficollis, Fabr.	Pretoria.
Colotes, sp. ?	Pretoria.

Fam. BOSTRYCHIDÆ.

Sinoxylon conigerum, Gerst.   Pretoria.

HETEROMERA.

I fully expected to find many more species of this group than I did near Pretoria, though the number of individuals was in an inverse ratio to the number of species. Most are naturally found on the scanty herbage of the veld, but some, as *Eletica rufa* and *Zonitis eborina*, frequent the leaves of Asclepiads. *Mylabris ophthalmica* throughout the summer is common on the rose-bloom of the hedges, and, as I have previously remarked, *Mylabris transversalis* is very injurious to the cultivated roses of Natal. *Dichtha cubica* is probably a mountain species; I certainly only found it in a barren rocky mountain pass: whilst *Psammodes striatus* is common everywhere and doubtless falls a prey to the large Geodephaga; in the dry season the empty body-cases of the *Psammodes* are found strewn over the plains.

I have been able to determine these species by comparison with the fine collection of Mr. Fred. Bates, now contained in the British Museum; and I have also to thank Mr. Champion for his assistance. I describe one species of which I can find no record. Some species of the obscure genera *Strongylium, Lagria, Nemognatha,* and *Trigonopus* I have not ventured to identify.

INSECTA. 199

Fam. TENEBRIONIDÆ.

Zophosis angusticostis, Deyr.	Pretoria.
Zophosis punctulata, Oliv.	Pretoria.
Himatismus buprestoides, Gerst.	Durban, *Natal*.
Machla porcella, Fåhr.	Pretoria.
Psammodes pierreti, Amyot.	Zoutpansberg.
Psammodes striatus, Fabr.	Pretoria.
Psammodes scabriusculus, Haag.	Pretoria.
Dichtha cubica, Guér.	Pretoria.
Amiantus gibbosus, Fåhr.	Pretoria.
Amiantus undosus, sp. n.	Pretoria.
Trachynotus angulatus, Fåhr.	Pretoria.
Trachynotus, sp.?	Pretoria.
Trigonopus, sp.?	Pretoria.
Anomalipes variolosus, Sol.	Pretoria.
Anomalipes intermedius, Dej.	Pretoria.
Anomalipes talpa, Fabr.	Pretoria.
Anomalipes complanatus, F. Bates.	Pretoria.
Micranterus validus, Boh.	Pretoria.
Strongylium, sp.?	Pretoria.
Lagria, sp.?	Pretoria.
Mylabris ophthalmica, Dej.	Pretoria.
Mylabris transversalis, Mars.	*Richmond Road, Natal*.
Mylabris lunata, Pall.	Zoutpansberg, Waterberg, Pretoria.
Mylabris mixta, Mars.	Pretoria and *Durban, Natal*.
Mylabris capensis, Linn.	Pretoria.
Mylabris tristigma, Gerst.	*Richmond Road, Natal*.
Mylabris gröndali, Billb.	Durban, *Natal*.
Eletica rufa, Fabr.	Pretoria.
Zonitis eborina, Fabr.	Pretoria.
Nemognatha, sp.?	Pretoria.

Description.

Amiantus undosus, sp. n. (Tab. I. fig. 1.)

Black; legs dark fuscous. Pronotum small, subglobular, strongly and reticulately rugose; head with the front rugose, but much more finely so than the pronotum; antennæ dark fuscous, blackish towards the base. Abdomen above subovate, slightly gibbous, strongly depressed posteriorly; lateral margins of the elytra convex; surface of the elytra covered with strong, waved, undulating rugosities and coarsely punctate, but not quite extending to their apices. Sternum very coarsely punctate and subrugulose; abdomen beneath glabrous, shining black, somewhat coarsely punctate.

Long. 12 millim. Pretoria.

Allied to *A. undatus*, Haag., but differing by the distinct pattern and sculpturing of the elytra. (*W. L. D.*)

Fam. CURCULIONIDÆ.

The most distinctive South-African genus of this family is *Brachycerus*, which in this region, at least, finds its headquarters. In and around Pretoria I found the species usually terrestrial, sometimes under stones, and frequently wandering among broken pieces of quartzite on hill-sides. The habits of *Polyclaeis equestris* and *cinereis* have already been referred to (*ante*, pp. 51-5). The weevils of the Old World are still so unworked by competent coleopterists that I have been unable to identify many species, but Mr. F. P. Pascoe has aided me considerably.

Proscephaladeres punctifrons, Boh.	Durban, Natal.
Proscephaladeres obesus, Boh.	Durban, Natal.
Polyclaeis equestris, Boh.	Pretoria.
Polyclaeis cinereis, Boh.	Pretoria.
Brachycerus apterus, Linn.	Pretoria.
Brachycerus cancellatus, Gylh.	Pretoria.
Brachycerus natalensis, Thm.	Pretoria.
Brachycerus, spp.?	Pretoria.
Hipporhinus pilularius, Fabr.	Pretoria.
Hipporhinus cornutus, Boh.	Pretoria.
Hipporhinus corniculatus, Thm.	Pretoria.
Hipporhinus, sp.?	Pretoria.
Cleonus, spp.?	Pretoria.
Lixus, spp.?	Pretoria.
Alcides senex, Sahlb., var.	Pretoria.
Acanthorrhinus dregei, Gylh.	Pretoria.
Sphenophorus, sp.?	Pretoria.

LONGICORNIA.

The absence of woods and forests on the "high veld," which may almost be said to compose the district of Pretoria, renders the number of longicorn beetles to be found but few in number and not particularly striking in size or appearance. Some I only met with in most unlikely places, such as *Compsomera elegantissima* in the billiard-room of the hotel at which I boarded, and *Taurotagus klugii* seen but once, and then on some wooden packing-cases in the town of Pretoria. *Anubis mellyi* swarms on the flowers (principally on *Scabiosa*, sp.) growing on the open veld, and *Promeces viridis* is also common on the same scanty flora. One of the most interesting species is *Amphidesmus analis*, which is generally found on leaves, where the females wonderfully resemble representatives of the genus *Lycus*; the long antennæ of the male render the deception less complete. The most showy species of the fauna, *Philagathes lætus*, is very abundant at the commencement of the summer in gardens, but is only on the wing for a short time.

Although I only found twenty-four species, three have proved to be undescribed, and my best thanks are due to Mr. C. J. Gahan, of the British Museum, without whose aid I should not have ventured to diagnose the new species.

Fam. PRIONIDÆ.

Tithoes confinis, Castel.	Pretoria.

Fam. CERAMBYCIDÆ.

Xystrocera globosa, Oliv.	Pretoria.
Taurotagus klugii, Lacord.	Pretoria.
Compsomera elegantissima, White.	Pretoria.
Phyllocnema latipes, De Geer.	Pretoria.
Anubis mellyi, White.	Pretoria.
Litopus dispar, Thoms.	Pretoria.
Promeces viridis, Pasc.	Waterberg and Pretoria.
Euporus callichromoides, Pasc.	*Durban, Natal.*
Paroeme gahani, sp. n.	Pretoria.
Clytanthus capensis, Lap. & Gory.	Pretoria.
Amphidesmus analis, Oliv.	Pretoria.
Philagathes lætus, Thoms.	Pretoria.

Fam. LAMIIDÆ.

Tragocephala sulphurata, sp. n.	Pretoria.
Ceroplesis bicincta, Fabr.	Pretoria.
Ceroplesis capensis, Linn., var. n.	Pretoria.
Ceroplesis brachyptera, Thoms.	Zoutpansberg, Waterberg, and Pretoria.
Phryneta spinator, Fabr.	Pretoria.
Rhaphidopsis zonaria.	Durban, Natal.
Tragiscoschema amabilis, Perr.	Durban, Natal.
Crossotus klugii, sp. n.	Pretoria.
Tetradia fasciatocollis, Thoms.	Pretoria.
Hecyrida terrea, Bertol.	Pretoria.
Morægamus globiceps, Har.	Pretoria.

Notes and Descriptions.

Paroeme gahani, sp. n. (Tab. I. fig. 7.)

Body brownish testaceous; eyes black; legs ochraceous, the apex of anterior femora, the apical halves of intermediate and posterior femora, apices of tibiæ, and apices of tarsal joints dark castaneous; apex of basal joint of antennæ broadly blackish, apices of second, third, fourth, and fifth joints narrowing, slightly darker. The basal joint of the antennæ is coarsely punctate above and abruptly truncate at apex; pronotum subnodulose and very sparingly pilose; the clytra are thickly and coarsely punctate.

Long. 15 millim. (*W. L. D.*)

Tragocephala sulphurata, sp. n. (Tab. I. fig. 9.)

Body sulphureous. Antennæ black. Head with a transverse fascia between antennal bases; eyes and a large triangular spot at their base black; pronotum with two slightly waved longitudinal fasciæ, fused and meeting on anterior margin, and apices of lateral tubercles black; margins of scutellum and elytral sutures, the last obliterated a little beyond centre and at apex, black; elytra ornamented with variable black spots (in some specimens being fasciate). Tarsi brownish beneath.

Long. 18 millim.

This species is closely allied to *T. variegata*, Bertol., but is much smaller in size and paler in hue. I found it on the open veld amongst dwarf flowering plants. (*W. L. D.*)

Ceroplesis capensis, Linn., var. n.

Cerambyx capensis, Linn. Syst. Nat. ed. xii. p. 628 (1767).

In this variety the red markings are almost entirely obliterated and the colour uniformly blackish brown.

Crossotus klugii, sp. n. (Tab. I. fig. 8.)

Crossotus sexpunctata, Klug, MS. Dej. Cat. ed. iii. p. 370.

Greyish brown mottled with darker coloration and darkly and coarsely punctate. Antennæ finely pilose above, strongly hirsute beneath, dark brownish, the joints (excluding basal) greyish at their bases. Head with the front pale, marked by some dark brown punctures, a black line behind base of antennæ and a black basal fascia; eyes black. Pronotum with some coarse, dark, scattered punctures, a distinct fuscous transverse fascia near anterior margin; four well-developed discal tubercles, with two central subobsolete ones between them, and two on each lateral margin, the posterior ones much the longest. Scutellum brownish grey, with a broad, central, blackish fascia. Elytra thickly, darkly, and very coarsely punctate, with three black tubercles on each side in longitudinal series, the anterior smallest; the central suture margined with small obscure greyish spots. Body beneath and legs somewhat paler in hue, coarsely punctured with brown; apices of the tibiæ and the tarsi brown.

Long. 12 to 14 millim.

Allied to *C. albicollis*, Guér., from West Africa, but the antennæ much less hairy.

I found this species on the stems of acacia trees, to which their colour-markings gave them great protective resemblance.

(*W. L. D.*)

EUPODA.

In the immediate vicinity of Pretoria scrubby woods are scarce, and consequently the numbers of Eupoda to be found are very limited. On the bare plains the Asclepiads attract many, especially such as *Euryope terminalis* and *Corynodes compressicornis*. Under stones in the dry season I have found *Aulacophora vinula*. Representatives of the Cryptocephalidæ were always obtained by beating trees. *Chrysomela opulenta* is a very common species.

I have to offer my best thanks to the following specialists:— Mr. Martin Jacoby for assistance in the Chrysomelidæ, and Mr. C. J. Gahan in the Galerucidæ, both of whom have here described species; whilst Mr. O. Janson has compared my Cassididæ with the fine collection of the late Mr. Baly, which is now in his possession.

Fam. CRIOCERIDÆ.

Crioceris puncticollis, Lac.	Durban, Natal.
Crioceris constricticollis, Clark.	Durban, Natal.

Fam. CRYPTOCEPHALIDÆ.

Gynandrophthalma anisogramma, Lac., var.	Pretoria.
Clythra wahlbergi, Lac.	Pretoria.
Camptolenes cribraria, Lac.	Pretoria.
Antipus rufus, De Geer.	Durban, Natal.
Cryptocephalus pustulatus, Fabr.	Pretoria.
Cryptocephalus dregei, Boh.	Pretoria.
Cryptocephalus decemnotatus, Suffr.	Pretoria.
Cryptocephalus pardalis, Suffr.	Pretoria.
Halitonoma epistomata, Fabr.	Pretoria.
Achænops facialis, sp. n.	Pretoria.

Description.

Achænops facialis, sp. n. (Tab. I. fig. 4.)

Below black; head black at the vertex, the lower part fulvous, deeply excavated; thorax fulvous, transverse, closely and finely punctured; elytra flavous, punctate-striate; the interstices finely punctured; the shoulders with a black spot; tibiæ fulvous at the base.

Length 1½ line.

Of cylindrical shape; the head broad and flat, the vertex black, closely and strongly punctured; the eyes very widely separated, but slightly emarginate; the lower portion of the face fulvous, very deeply excavated, the excavation bounded at the sides by a sharp edge; antennæ short, black, the lower five joints fulvous, the terminal joints gradually widened; thorax about twice as broad as long, subcylindrical, the lateral margins straight at the base, rounded in front, the surface very closely punctured, fulvous, the extreme basal margin black; scutellum black, its apex truncate; elytra cylindrical, flavous, the extreme basal margin and a spot at the shoulder black, the disk distinctly punctate-striate; the interstices very finely punctured and here and there transversely wrinkled; pygidium piceous, strongly punctured; underside and the femora black; the coxæ, base of the femora and that of the tibiæ fulvous; prosternum with a triangular tooth at the middle of the basal margin, fulvous.

A single specimen, differing from the other known species by the different coloration and the very deep facial excavation.

(*M. Jacoby.*)

Fam. EUMOLPIDÆ.

Colasposoma pubescens, Lefèvre.	Pretoria.
Euryope terminalis, Baly.	Pretoria.
Calomorpha wahlbergi, Stål.	Durban, Natal.
Pseudocolaspis sericata, Marsh.	Pretoria.
Corynodes compressicornis, Fabr.	Pretoria.
Menius distanti, sp. n.	Pretoria.

Description.

Menius distanti, sp. n. (Tab. I. fig. 6.)

Piceous; the basal and apical joints of the antennæ and the legs fulvous; head and thorax not very closely and strongly punctured; elytra dark or pale fulvous, regularly and strongly punctate-striate.

Length 1½ line.

Head strongly punctured, the vertex piceous, deeply sulcate above the eyes, the lower portion fulvous; antennæ only extending to the base of the elytra, fulvous, the eighth and ninth joints piceous, the second joint slightly longer and thicker than the third; thorax more than twice as broad as long, widened at the middle, the sides rounded, the surface deeply but not very closely punctured, leaving a smooth central longitudinal space; elytra slightly depressed below the base, rather lighter in colour than the thorax, deeply and regularly punctate-striate, the interstices impunctate but convex at the sides, the shoulders prominent : underside piceous; legs fulvous; prosternum longer than broad, slightly narrowed between the coxæ.

The two specimens obtained differ slightly in the colour of the elytra, which are much paler in one than in the other; the species is allied to *M. chalceatus*, Lefèvre, but differs in the more strongly punctured thorax, the flat interstices of the elytra (the sides excepted), and in the fulvous legs. (*M. Jacoby.*)

Fam. CHRYSOMELIDÆ.

Chrysomela opulenta, Reiche.	Pretoria.
Polysticta clarkii, Baly.	Pretoria; *Durban, Natal*.
Podontia nigrotessellata, Baly.	Pretoria.

Fam. GALERUCIDÆ.

Aulacophora vinula, Erichs.	Pretoria.
Hyperacantha oculata, Karsch.	Pretoria.
Asbecesta cyanipennis, Harold.	Zoutpansberg.
Sphæroderma indica, Fabr.	Pretoria.
Ænidea pretoriæ, sp. n.	Pretoria.
Monolepta flaveola, Gerst.	*Durban, Natal.*
Spilocephalus viridipennis, Jac.	Pretoria.
Ootheca modesta, sp. n.	Pretoria.

Notes and Descriptions.
(By C. J. GAHAN, M.A., F.E.S.)

Ootheca modesta, sp. n. (Tab. I. fig. 11.)

Testaceous; underside of body (abdomen and sides of prothorax excepted), legs, and scutellum black. Sides of prothorax slightly diverging from the base up to the anterior third, thence converging to the apex; pronotum convex, minutely and rather thickly punctured, its greatest width about twice the median length. Elytra minutely and very closely punctured. Abdomen fuscous-testaceous. Underside of body and legs thinly clothed with short grey hairs. Antennæ a little longer than half the body, the third joint about equal to the fifth, the fourth slightly longer, the sixth and following joints subequal or scarcely perceptibly diminishing in length, each shorter than the fifth; each of the joints from the third to the tenth slightly thickened towards the apex.

Epipleures of elytra moderately broad in front, gradually narrowed posteriorly, and entirely disappearing just beyond the middle. Tibiæ unarmed. First joint of posterior tarsi equal in length to the two succeeding joints united. Anterior cotyloid cavities closed in behind.

Long. 5½ millim.

This species is smaller than *O. mutabilis*, Sahlb. (Peric. Entom. Species Insect. (1823) p. 64, pl. 3. figs. 8–10), the prothorax is less rounded at the sides and the whole insect less ovate in form; but it agrees with that species in having short elytral epipleures, closed anterior cotyloid cavities, appendiculate claws.

So that, on the whole (considering *O. mutabilis*, Sahlb., as the type of the genus), the present species seems best placed in *Ootheca*.

Chapuis, in his characterization of this genus ('Genera des Coléoptères,' xi. p. 173), has stated that the anterior cotyloid cavities are open behind. But this statement cannot be accepted as correct, unless Chapuis was mistaken in his identification of *Crioceris mutabilis*, Sahlb., the species which he names as the type of his genus. (*C. J. G.*)

Spilocephalus viridipennis, Jacoby. (Tab. I. fig. 12.)

Spilocephalus viridipennis, Jacoby, Trans. Ent. Soc. Lond. 1888, p. 202, pl. vii. fig. 12 (♀ ?).

Mr. Distant has taken one male specimen which I refer with some doubt to this species. This specimen not only differs from the type by certain well-marked characters which I consider to be sexual, but disagrees also in some minor details with Mr. Jacoby's description; so that it may possibly represent a distinct species. The sides of the prothorax slightly diverge from the base up to about the anterior third, and thence converge to the apex; the disk of the prothorax is almost impunctate; the greatest width of the thorax is rather less than twice the length. The punctuation of the elytra is arranged in double rows, the intervals between each pair of rows forming slightly raised longitudinal lines, which are more distinct on the disk.

The antennæ of the male are almost as long as the body; the hind tibiæ of the same sex are each furnished with a long slender spur, which arises on the inner side at about the beginning of the distal fourth, and passing alongside the tibia, terminates just beyond it in a little spine. The apex of the last ventral segment of the male has a slight notch or incision on each side, a short median lobe being thus formed.

In other respects (the characters here given excepted) Mr. Distant's specimen agrees with Jacoby's description.

The remarkable tibial spur, which, like the tibiæ, is provided with short grey hairs, except on the terminal spine, is possessed also by the males of *Xenarthra?* *calcarata*, Gerstaeck., *Xenarthra bipunctata*, Allard (Ann. Ent. Belg. 1889, p. cxiv), and a species* from the Transvaal hitherto undescribed, which I name in honour of my distinguished friend. These three species, though differing from the male described above by the greater length of the

---

* *Spilocephalus distanti*, sp. n.

♂. Head, legs, and antennæ fulvous, the latter slightly infuscate towards the apex. Prothorax, elytra, and underside of body metallic green. Head deeply excavated in front, with a median process passing upwards from the epistome; vertex transversely impressed between the eyes, and with a short triangular lobe, impressed longitudinally, projecting between the antennæ. Prothorax with its sides gradually diverging from the base almost up to the apex, its anterior width about twice the median length; the disk dull, with a faint, median, longitudinal, impressed line, and with two shallow depressions—one on each side near the base. Elytra rather strongly and very closely punctured, subnitid. Antennæ a third longer than the body. Hind tibiæ each with a long slender spur.

Hab. Transvaal. (Brit. Mus. Collection.) (*C. J. G.*)

antennæ (a third, at least, longer than the body) and by the excavation of the front of the head, will probably have to rank in the ame genus.

Ænidea? pretoriæ, sp. n. (Tab. I. fig. 10.)

Fulvo-testaceous; head above, scutellum, body underneath, and coxæ black; last four or six joints of the antennæ fuscous. Face rather short, lower margin of clypeus transversely raised; vertex impressed transversely between the eyes, and longitudinally between the antennæ. Sides of prothorax slightly diverging from the base to the anterior third, thence converging; pronotum nitid and impunctate, with a transverse depression just behind the middle. Elytra somewhat elongate, closely but not distinctly punctured; apices somewhat sharply rounded; epipleures prolonged up to the apical border. Tibiæ unarmed; first joint of hind tarsi almost as long as the three succeeding joints combined. Antennæ (♀?) almost as long as the body; scape slightly curved, third and fifth joints subequal, fourth perceptibly longer, sixth and succeeding joints gradually and very slightly diminishing in length.

Fam. CASSIDIDÆ.

Basipta stolida, Boh.	Durban, *Natal*.
Cassida punctata, Fabr.	Durban, *Natal*.
Cassida scripta, Fabr.	Pretoria.
Cassida hybrida, Boh.	Waterberg.
Cassida lurida, Boh.	Zoutpansberg.

Fam. COCCINELLIDÆ.

Adalia flavomaculata, De Geer.	Pretoria.
Cydonia lunata, Fabr.	Pretoria.
Cydonia quadrilineata, Muls.	Pretoria.
Alesia inclusa, Muls.	Pretoria.
Exochomus nigromaculatus, Goeze.	Pretoria.
Epilachna dregei, Muls.	Pretoria and *Durban, Natal*.
Epilachna hirta, Thunb.	Durban, *Natal*.
Epilachna bifasciata, Fabr.	Pretoria.

Fam. ENDOMYCHIDÆ.

Ancylopus fuscipennis, sp. n.   (Tab. IV. fig. 10.)

Head and prothorax reddish brown, elytra dark brown. Pronotum slightly transverse, thickly and minutely punctulate, with a short longitudinal impression on each side near the base; sides of the pronotum very slightly diverging from the base to a little beyond the middle, thence converging anteriorly; anterior margin arcuately emarginate. Elytra thickly and minutely punctulate. Legs and underside dark brown, with the bases of the femora and the middle of the breast somewhat reddish. Long. 5 millim.   Pretoria.

This species appears to be most nearly allied to *Ancylopus bivittatus*, Perch., which it closely resembles in form and punctuation, but from which it is to be distinguished by its smaller size and the dark brown coloration extending over the whole of the elytra.   (C. J. Gahan.)

HYMENOPTERA.

I am indebted to Mons. Henri de Saussure for having worked out my collections in this order. The types of the new species which he has described are now in his fine collection at Geneva.

Fam. APIDÆ.

Xylocopa inconstans, Smith.	Pretoria.
Megachile maxillosa, Guér.	Pretoria.
Helioryctes melanopyrus, Smith.	Pretoria.

Fam. VESPIDÆ.

Eumenes tinctor, Chr.	Pretoria.
Synagris mirabilis, Guér.	Pretoria.
Pterochilus insignis, Sauss.	Pretoria.
Belonogaster rufipennis, De Geer.	Pretoria.

Fam. SPHEGIDÆ.

Larra ornata, Lep.	Pretoria.
Chlorion xanthocerum, Meig., var.	Pretoria.
Sphex nigripes, Sm., var.	Pretoria.
Pelopæus spirifex, Linn.	Pretoria.
Ammophila bonæ spei, Lep.	Pretoria.
Ampulex nigro-cærulea, sp. n.	Pretoria.

Fam. POMPILIDÆ.

Homonotus cærulans, sp. n.	Pretoria.
Homonotus pedestris, sp. n.	Pretoria.
Priocnemis hirsutus, sp. n.	Pretoria.
Cyphononyx antennata, sp. n.	Pretoria.
Mygnimia belzebuth, sp. n.	Pretoria
Mygnimia depressa, sp. n.	Pretoria.
Mygnimia distanti, sp. n.	Pretoria.
Mygnimia fallax, sp. n.	Pretoria.

Fam. SCOLIIDÆ.

Discolia caffra, Sauss.	Pretoria.
Discolia præcana, sp. n.	Pretoria.
Discolia præstabilis, sp. n.	Pretoria.
Elis barbata, Sauss.	Pretoria.
Mesa diapherogamia, sp. n.	Pretoria.

Fam. MUTILLIDÆ.

Mutilla albistyla, sp. n.	Pretoria.
Mutilla tetensis, Gerst.	Pretoria.

Fam. FORMICIDÆ.

Camponotus grandidieri, Forel.	Pretoria.
Carebara vidua, Linn.	Pretoria.
Dorylus helvolus, Linn.	Pretoria.

Fam. CHRYSIDIDÆ.

Chrysis (Pyria) lynica, Fabr.	Pretoria.

Fam. ICHNEUMONIDÆ.

Hemipimpla caffra, sp. n.	Pretoria.
Hemipimpla calliptera, sp. n.	Pretoria.
Distantella trinotata, gen. et sp. n.	Pretoria.

Fam. BRACONIDÆ.

Bracon flagrator, Gerst.	Pretoria.
Bracon fastidiator, Fabr.	Pretoria.

Fam. TENTHREDINIDÆ.

Athalia bicolor, sp. n.	Pretoria.

Descriptions.

(Par M. Henri de Saussure.)

Familia SPHEGIDÆ.

Genus *Ampulex*, Jurine.

Ampulex nigrocærulea, Saussure (sp. n.). (Tab. IV. fig. 6.)

Sat minuta, nigra, lævigata, mandibulis et clypei apice rufis. Pronotum haud sulcatum, postice tuberculo acuto instructum. Metathorax abdominisque segmenta 1^m et 2^m splendide nigrocærulea; metanoto bidentato, etsi transverse carinulato; abdominis segmenta 2^m et 3^m cinereo-pubescentia. Alæ hyalinæ, basi inquinatæ, ultra medium vitta transversali nigra.

♀. Long. 15 mill., al. 9.

Tête, prothorax, mésothorax, écusson et postécusson, noirs. Tête assez grosse, à bord postérieur arqué, semée de petites ponctuations. Mandibules, extrémité du chaperon et scape des antennes en dessous, roux. Le front lisse, n'offrant que deux faibles et courtes carinules surantennaires.—Pronotum lisse, non partagé; son tubercule aigu; ses bords latéraux en carènes arquées mousses; sa partie antérieure finement striolée en longueur et un peu en éventail (en forme d'arbuste). Le lobe supérieur des propleures lisse comme le reste, sans aucune ponctuation. Mésonotum semé d'assez faibles ponctuations; mésopleures plus fortement ponctuées. Métathorax d'un beau bleu métallique sombre; ses angles formant deux dents triangulaires aiguës, comprimées et arquées, aussi fortes ou plus longues que chez l'*A. compressa*. La sculpture du métanotum presque la même que chez la *sibirica*; la 3e bande intercarinaire de chaque côté, s'étendant jusqu'à la base, aussi large que la 4e; celle-ci un peu rétrécie en arrière. Le bord postérieur un peu sinué de chaque côté, comme chez la *compressa*. La plaque postérieure plate, grossièrement réticuleusement ponctuée.—Abdomen noir; ses deux premiers segments d'un bleu d'acier noirâtre, semés de petites ponctuations; le 2e sur ses côtés et sur son bord postérieur et le 3e, revêtus d'un fin duvet gris. Les suivants glabres; l'anus un peu roussâtre.—Pattes noires; tarses roussâtres en dessous.—Ailes hyalines à nervures noires, avec une bande brune qui part du stigma et de la radiale; les cellules de la base et la cellule anale salies de brun; la 3ᵉ cellule cubitale rétrécie de ¼ vers la radiale. Ailes postérieures hyalines.

Pretoria, 1 ♀.

Espèce offrant sensiblement les mêmes formes que l'*A. compressa*, F.

Familia Pompilidæ.

Genus *Homonotus*, Dahlbom, Saussure.

Homonotus cærulans, Saussure (sp. n.). (Tab. V. fig. 1.)

Niger, cærulans; alis fuscis, violascentibus. Caput compressum, postice concavum; clypeo grandi, plano, late bilobato; occipite acuto, arcuato. Thorax anterius subdepressus; pronoto, mesonoto æquilongo; metathorace truncato, strigato. Abdomen sessile, ovato-conicum. Ungues dente obliquo armati.

♀. Long. 17 mill., al. 11·5.

♀. Antennes noires à reflets roussâtres dans leur seconde moitié. Tête plus large que le thorax, lisse, noire, avec des reflets bleuâtres au sommet en devant, fortement concave en arrière; l'occiput tranchant, subbisinué au milieu, passant derrière les yeux. Ocelles aplatis, logés contre des dépressions irrégulières. Yeux étroits, parallèles, à bord interne droit. Fossettes antennaires grandes et profondes, réunies en dessous en une gouttière bisinuée, faisant paraître le chaperon en relief. Celui-ci très grand, plat et lisse, formant une grande lame prolongée en bas, à bords latéraux parallèles, à bord inférieur échancré au milieu et largement bilobé; ses deux pointes latérales obliques, atteignant plus haut que bas des yeux, séparées du prolongement par une petite échancrure.

Thorax long et étroit, lisse; de profil à peine convexe en dessus. Pronotum aplati, un peu rétréci en avant, fortement échancré en arrière d'une manière presque angulaire; ses bords arrondis; ses côtés creusés d'une gouttière en V; leur angle inférieur placé en arrière; leur bord postérieur peu oblique, peu sinué. Mésonotum à peine aussi long que le pronotum, avec deux sillons très latéraux. Écailles alaires petites, bordées de roux. Écusson aplati. Métathorax point déprimé; en dessus un peu moins long que large, partagé par un sillon, fortement ridé en arrière, obsolètement en avant, offrant un sillon en avant de chaque stigmate ; le sillon transverse de la base large, touchant le postécusson ; la plaque postérieure ridée, à bords arrondis ; métapleures veloutés, offrant souvent une petite arête oblique.—Abdomen comprimé dès le 3e segment ; ses 2 premiers segments un peu aplatis.—Pattes médiocres. Fémurs tous presque également forts ; tibias armés seulement de très petites épines. Tarses garnis en dessous de deux rangées de fortes spinules. Griffes bifides, mais leur dent inférieure beaucoup plus courte que la supérieure et tronquée.—Ailes étroites. La cellule radiale assez large, aussi aiguë à sa base qu'à son extrémité, à bord postérieur arqué jusqu'à la 3e veine transverso-cubitale, puis droit et oblique, formant à la cellule une pointe courte et aiguë. Les 2e et 3e cellules cubitales plus larges que hautes ; la 2e moins grande, en carré oblique, recevant la 1e veine récurrente à son 2e tiers. Aux ailes postérieures la veine anale s'insérant un peu après le point d'origine de la veine cubitale.

Pretoria, 1 ♀.

Cette espèce est très caractérisée par ses formes qui rappellent celles de certains *Salius*, mais le métathorax n'est pas échancré comme dans ce genre.

Homonotus pedestris, Saussure (sp. n.). (Tab. V. fig. 6.)

Gracilis, niger, cinereo-sericans ; vertice acuto, transverso ; clypeo haud producto ; pronoto postice angulatim inciso ; metathorace lævigato, subtruncato, postice erecto-fusco-setoso ; abdomine prismatico, segmentis basi plumbeatis, ultimo albo ; pedibus gracilibus, tibiis anticis brevissimis, reliquis longe remote spinosis ; femoribus et tibiis posticis rufis nigro-spinosis, femoribus basi et apice nigris ; calcaribus tibiarum rufis ; alis pallide fuscis, 3^a areola cubitali petiolata.

♂. Long. 12 mill., al. 10.

♂. Très grêle. Antennes noires, noueuses, les articles étant renflés en dedans vers leur base, comme s'ils ne se faisaient pas suite ; leur premier article aplati en dessous et blanc ; les articles 3e-5e roux en dessous. Tête comprimée, plate en arrière ; le vertex formant une arête vive droite d'un œil à l'autre, la tête n'étant point renflée en arrière des yeux. Ocelles assez grands, les postérieurs placés presque contre l'arête du vertex. Front subconvexe, partagé par un sillon. Yeux légèrement sinués à leur bord interne, un peu divergents vers le haut, n'atteignant pas la base des mandibules. Chaperon transversal, sinué dans toute sa largeur; le labre grand et arrondi. Bouche noire; palpes médiocres.

Thorax comprimé rétréci en arrière, voûté en avant. Pronotum aussi long que le mésonotum, échancré à angle obtus en arrière, rétréci en avant, à bords arrondis; ses lobes latéraux hauts, excavés, terminés d'une manière presque carrée, à angles arrondis; leur bord antérieur peu oblique; leur bord postérieur oblique, sinué; leur angle sous-alaire arrondi, non tuberculé. Mésonotum uni. Écailles bordées de roux. Écusson ayant sa partie saillante étroite, allongée, aplatie. Métanotum un peu oblique, d'un poli mat, subconvexe, tronqué d'une manière peu franche; sa face postérieure peu haute, subconvexe, garnie de gros poils courts bruns ou à reflets blancs, relevés de bas en haut.—Abdomen très grêle, sessile, comprimé en dessous. Les segments d'un gris plombé avec leur bord postérieur plus ou moins largement noir (nud); segments 5e et 6e avec pubescence grise en dessus; le 7e blanc.—Pattes grêles. Fémurs antérieurs très grêles, armés d'une épine apicale interne; les suivants plus larges, comprimés; ceux de la 3e paire avec 2 : 3 spinules en dessus vers leur extrémité. Tibias antérieurs très courts, pas plus longs que le métatarse, armés au bord externe de 2 épines et d'une 3e apicale très longue; offrant en dessus 2–3 rangées de très petites épines, et 3 épines apicales, dont 2 très petites et l'interne longue. Tibias et métatarses des 2e et 3e paires armés en dessus et en dehors de deux rangées d'épines, très espacées et peu nombreuses (2 : 3 ou 3 : 4). Fémurs postérieurs roux avec la base noire; leur extrémité et celle des tibias avec un peu de noir. Toutes les épines noires, mais les éperons des tibias roux. Griffes bifides.

Pretoria, 1 ♂.

Espèce très voisine de l'*H. ibex*, Sauss., de Madagascar, mais dont tous les tibias sont noirs.

Genus *Priocnemis*, Dahlb.

Priocnemis hirsutus, Saussure (sp. n.). (Tab. V. fig. 3.)

Niger, fulvo-hirsutus; antennis, ore, clypeo, orbitis internis prothorace, tegulis pedibusque, aurantiis; abdomine depressiusculo, sessili; segmentis 1°–3° late flavo-aurantio limbatis, reliquis aurantiis; alis fusco-violaceis.

♀. Long. 18 mill., al. 16·5.

Tête très finement chagrinée; chaperon convexe, ponctué. Thorax long, comprimé; le pronotum très court velouté, largement sinué en arrière; le milieu des bords latéraux offrant souvent un petit tubercule. Mésonotum un peu tricaréné; les deux dépressions entre les carinules souvent obliquement striées et la partie antérieure striée en longueur et ponctuée (chez certains individus le mésonotum est lisse). La partie saillante de l'écusson triangulaire, n'atteignant pas le bord postérieur, ponctué-strié ou lisse. Postécusson lisse, un peu renflé en tubercule arrondi au milieu en arrière. Métathorax tronqué à arête très arrondie; ses angles très arrondis; sa base sans tubercles; sa face supérieure convexe transversalement, assez finement striée; la base en avant du sillon transversal lisse et fortement partagée; la face postérieure plate ou subconvexe. Métapleures lisses. Tout le métathorax hérissé de longs poils fauves.—Abdomen un peu déprimé, sessile, elliptique. Les larges bandes jaunes des 3 premiers segments échancrées au milieu.—Pattes armées de petites épines; tarses portant en dessous deux rangées de spinules espacées. Hanches souvent noires au bord externe. Griffes armées d'une très petite dent.—Ailes d'un brun foncé, avec leur extrême base et la base de la côte roux-veloutés. La 2ᵉ veinule transversocubitale presque droite, non courbée en crochet en arrière; la 2ᵉ cellule cubitale presque 3 fois plus large que haute, recevant la 1ᵉ veine récurrente très près de son extrémité. Aux ailes postérieures la veine anale s'insérant bien au-delà du point d'origine de la veine cubitale.

Pretoria, 1 ♀.

Je possède d'autres individus provenant du Cap de Bonne Espérance.

Ce *Priocnemis* est par sa vénulation alaire voisin des *Mygnimia*. Il est remarquable par son corps velouté et hérissé de poils gris-fauves, surtout longs au métathorax.

Genus *Cyphononyx*, Dahlb.

Cyphononyx antennata, Saussure (sp. n.). (Tab. V. fig. 2.)

Niger, velutinus, nigro-hirsutus; antennis, clypeo, ore, genubus, tibiis et tarsis, aurantiis; orbitis obscure rufis; alis nigro-violaceis. Postscutellum compressum, truncatum; metanotum arcuatum, strigatum, utrinque carinatum, basi utrinque tuberculatum. Tibiæ posticæ ♀ serrulatæ.

♀. Long. 19 mill., al. 17. ♂. Long. 15 mill., al. 14.

♀. Insecte velouté, finement ponctué ou rugulé. Chaperon transversal. La face aplatie un peu excavée, bordée le long des yeux pas des bourrelets très obsolètes qui se rejoignent en arc de cercle ou en échancrure de cœur en sommet de la face. Un sillon descendant de l'ocelle antérieur se prolonge jusqu'au chaperon, partageant la protubérance surantennaire. Orbites postérieures et internes vers le bas, parfois toute la face et la tête par derrière, d'un roux sombre. Pronotum formant de chaque côté un tubercule transversal arrondi. Écusson ayant sa partie saillante triangulaire étroite, avec une tache rousse avant l'extrémité. Postécusson comprimé en tubercule ou en dos d'âne arrondi; tronqué-arrondi en arrière. Métathorax convexe d'avant en arrière, de profil descendant par une courbe, non tronqué; assez plat transversalement, strié, formant de chaque côté une arête vive qui part en dedans du stigmate, et en dehors de laquelle est une gouttière latérale. La ligne médiane occupée par un sillon en gouttière obsolète. Le sillon transversal de la base linéaire, mais la partie qui le sépare du postécusson enfoncée, se confondant presque avec le sillon, finement striée en travers.—Abdomen fusiforme, sans reflets bleuâtres.—Pattes armées de très nombreuses épines, fortes et courtes, souvent brunes. Tibias postérieurs serrulés par dents écailleuses. Tarses avec deux rangées de fortes spinules en dessous; métatarses fort épineux. Griffes bifides; leur branche supérieure longuement droite, l'inférieure de moitié moins longue, pas plus large que la supérieure, à pointe mousse, non tronquée. Tarses des 2ᵉ et 3ᵉ paires ayant leurs articles un peu noirs à leur base. —Ailes d'un brun très foncé; la cellule radiale plus aiguë à sa base qu'à son extrémité; la 2ᵉ cubitale en trapèze, un peu plus large que haute, recevant la 1ᵉ veine récurrente près de son extrémité. Aux ailes postérieures la veine anale s'insérant sur le point d'origine de la veine cubitale.

♂. La face au dessus des antennes n'étant pas distinctement excavée et bordée. Tibias postérieurs non serrulés.

Pretoria, 2 ♀, 1 ♂.

Espèce très voisine des *C. dolosus* et *grandidieri*, Sauss., de Madagascar; s'en distinguant par sa face ♀ excavée, et par ses tibias et tarses jaunes.

Genus *Mygnimia*, Smith.

Mygnimia belzebuth, Saussure (sp. n.). (Tab. V. fig. 8.)

Gracilis, compressa, aterrima, velutina, nigro-hirsuta. Antennæ filiformes. Scutellum et postscutellum compressa. Metathorax truncatus, supra tenuiter strigatus, in medio compresso-rotundatus, angulis posticis rotundatis. Abdomen prismatico-fusiforme, sessile. Tibiæ et tarsi antici subtus rufescenti-sericantes. Alæ nigerrimæ, violaceæ.

♂. Long. 25 mill., al. 24.

♂. Antennes légèrement aplaties, non dilatées. Palpes bruns; le 3ᵉ article des maxillaires long, les 3 derniers courts et égaux. Pronotum faiblement sinué en arrière. Écusson et postécusson comprimés en dos d'âne mousse. Métathorax à peine bituberculé à sa base; ses tubercules très arrondis, obsolètes; le sillon transverse de la base profond; l'espace qui le précède non strié, fortement partagé, un peu convexe de chaque côté. Le métanotum un peu élevé en dos d'âne arrondi au milieu; la face postérieure peu élevée. Griffes armées d'une dent aiguë.— Ailes à beaux reflets bleus un peu violets; aux antérieures la 1ᵉ veine récurrente aboutissant un peu avant l'extrémité de la 2ᵉ cellule cubitale. Aux ailes postérieures la veine anale s'insérant à angle aigu bien au-delà du point d'origine de la veine cubitale.

Pretoria, 1 ♂.

Mygnimia depressa, Saussure (sp. n.).

Statura media ; nigra ; antennis, pedibus abdominisque ultimo segmento, aurantiis ; capite, prothorace, mesonoto, tegulis scutellique trigono supero, rufis, pilis rufo-aureis vestitis, velutinis ; metanoto strigato ; tibiis posticis ♀ serrulatis ; alis nigerrimis, violaceis, ima basi rufo-velutinis.
♀. Long. 25 mill., al. 22.

De taille assez petite pour le genre. Mandibules et palpes roux. Chaperon grand, non largement sinué, mais à bord inférieur arqué, échancré au milieu, les angles de l'échancrure formant comme deux petits lobes arrondis. Mésonotum avec 4 sillons ; les deux internes obsolètes, en gouttières. La face supérieure de l'écusson assez aplatie, en triangle aigu. Post-écusson lisse, formant au milieu en arrière presque un tubercule arrondi. Métathorax tronqué perpendiculairement en arrière, strié en travers, dépourvu de tubercules stigmataires ; le sillon transverse de sa base profond ; l'espace qui le sépare du post-écusson étroit, lisse, avec un gros point enfoncé au milieu. La face postérieure subconcave, striolée, sans aucun angle latéral ; son bord supérieur très arqué.—Abdomen déprimé, sessile, lisse et poli ; extrême bord des segments 4º et 5º roussâtre ; le 6º hérissé de poils roux.—Pattes armées d'épines courtes et fortes. Tibias postérieurs ♀ portant en dessus une carène serrulée de petites dents triangulaires.—Aux ailes antérieures la cellule radiale tronquée obliquement ; la 2º veinule transverso-cubitale peu courbée en arrière pour se continuer avec la 1ᵉ veine récurrente. Aux ailes postérieures la veine anale s'insérant sur la veine discoïdale bien au-delà du point d'origine de la veine cubitale ; l'une et l'autre fortement courbées en crochet à leur insertion.

Pretoria, 1 ♀.

Espèce voisine pour le livrée de la *M. nenitra,* Sauss., mais à formes plus déprimées ; s'en distinguant par son chaperon non largement sinué, son métathorax strié, par la vénulation des ailes, etc.—On trouve dans le midi de l'Afrique plusieurs espèces très voisines de la *M. depressa* :

M. peringueyi, n., à thorax comprimé, à chaperon transversal, largement sinué ; la 1ᵉ veine récurrente n'atteignant pas le bout de la 2ᵉ cellule cubitale (donc presque un *Priocnemis*).

M. hottentota, n., à métathorax fortement ridé, caréné de chaque côté, peu distinctement tronqué ; à l'aile postérieure la veine anale s'insérant sans crochet à peine au-delà de l'origine de la veine cubitale.

Mygnimia distanti, Saussure (sp. n.).   (Tab. V. fig. 7.)

Nigra ; antennis, pedibus, capite, ore, prothorace, vitta mesopleurûm, tegulis abdominisque ultimo segmento, flavoaurantiis ; clypeo sinuato ; metathorace toto strigato ; alis nigro-cæruleis.

♀. Long. 28 mill., al. 23.

Taille et formes comme chez la *M. depressa*. Le chaperon assez largement sinué, comme d'habitude. Thorax glabre, lisse et poli ; le mésonotum plat, n'offrant que deux fins sillons. Le reflet de l'écusson étroit et allongé, d'abord rapidement rétréci, puis presque parallèle, tronqué au bout. Postécusson offrant au milieu une petite bosse arrondie n'atteignant pas le bord postérieur ; ses parties latérales avec 3-4 rides transversales. Métanotum partout fortement strié, même sur l'aire postscutellaire ; sa base offrant de chaque côté un tubercule arrondi obsolète crénelé ; son extrémité tronquée perpendiculairement, à angle vif ; la plaque postérieure plate, plus finement striée ; son bord supérieur peu arqué, formant avec les arêtes latérales une sorte d'angle, ces arêtes crénelées au sommet. —Abdomen subdéprimé, lisse, sessile. Le 5ᵉ segment ayant son bord apical roux-testacé ; le 6ᵉ velouté d'un duvet orangé et hérissé de poils jaunes et bruns.—Pattes armées de nombreuses épines brunes, plus longues que chez la *M. depressa* ; tibias postérieurs ♀ carénés et serrulés comme chez cette espèce.

Ailes antérieures comme chez la *M. depressa*, mais avec la 2ᵉ veine transverso-cubitale plus fortement courbée en arrière ; aux ailes postérieures la veine anale s'insérant un peu au-delà du point d'origine de la veine cubitale et sans crochet (à angle aigu) ; la veine cubitale s'échappant en formant un grand arc en demi-cercle.

Pretoria, 1 ♀.

Espèce rappelant un peu le *Priocnemis clypeatus*, Klug (Symbol. Phys. pl. 39. fig. 14).

Mygnimia fallax, Saussure (sp. n.). (Tab. V. fig. 5.)

Minor, gracilis; antennis, capite, prothorace, mesopleuris, mesonoto, tegulis, scutello et postscutello in medio, pedibus, abdominisque segmentis 6° et 7° rufis vel aurantiis; metathorace rotundato, strigato; alis fusco-violaceis.

♂. Long. 14 mill., al. 12.

Tête et thorax soyeux-veloutés.

Chaperon un peu convexe, en trapèze renversé, à bord inférieur droit. Mésonotum offrant souvent de chaque côté une bande noire dans sa gouttière latérale, et le long du bord antérieur. Écusson convexe. Postécusson formant au milieu une bosse arrondie, un peu strié sur les côtés déprimés. Métathorax offrant de chaque côté un très faible tubercule stigmataire, point tronqué, mais arrondi, tombant obliquement en arrière, ridé en travers; sa base entre le sillon transverse et le postécusson lisse ou faiblement strié en travers.—Abdomen fusiforme, revêtu d'un duvet soyeux grisâtre, devenant roussâtre aux derniers segments; le 5e segment passant déjà au roussâtre.—Pattes grêles.—Ailes antérieures: la cellule radiale assez aiguë; la 2e veine transverso-cubitale fortement courbée en crochet en arrière. Aux ailes postérieures la vénulation comme chez la *M. transvaaliana*, mais la veine anale s'insérant très près du point d'origine de la veine cubitale.—*Var.* Mésonotum noir avec 3 bandes rousses.

Pretoria, 1 ♂.

J'aurais pris cet insecte pour le mâle de la *M. distanti* si son métathorax n'était pas arrondi au lieu d'être tronqué.

La *M. peringueyi* du Cap de Bonne Espérance est une espèce très voisine, mais plus grande, à postécusson tuberculé, et chez qui les veines anale et cubitale de l'aile postérieure s'insèrent sur le même point, et où la 1e veine récurrente s'insère sur la 2e cubitale avant l'extrémité de cette cellule (voir page 219).

Familia SCOLIIDÆ.

Genus *Scolia*, L.

Discolia præcana, Saussure (sp. n.). (Tab. IV. fig. 11.)

Tota nigra, sat fortiter punctata, albido-pilosa; thoracis et abdominis bascos pilis in certa luce fuscis vel nigris. Metathorax supra crassius punctatus, facie postica plana, punctata, margine supero leviter arcuato, acuto. Abdomen cærulescens; primo segmento sat brevi; reliquis albido-fimbriatis, lateribus abunde albido-hirsutis. Pygidium planatum, subexcavatum, rufescens, oblique sparse punctatum, apice rotundatum. Pedes albido-pilosi; tibiarum calcaria antica rufa, parum incurva; reliqua nigra, apice imo rufo. Ungues rufi, apice nigro. Alæ fuscæ, violascentes, vitta lata marginis antici nigra, limbo apicali a sinu anali subvitreo; areola radiali et linea transversa primæ cubitalis pallidioribus, flavicantibus; areola radiali truncata, secundam cubitalem paulum superans.

♂. Long. 15 mill., al. 13.

La tête est finement ponctuée; le vertex lisse; le thorax est plus finement ponctué en avant qu'en arrière; le métanotum l'étant le plus fortement. La cellule radiale est tronquée, à angle apical-postérieur arroudi, et dépasse un peu la 2ᵉ cubitale. Les ailes ont la même livrée, très caractéristique, que la *Liacos nigrita*; elles sont inégalement brunies à la base, subhyalines depuis le milieu, mais avec une bande noire couvrant les cellules cubitales, occupant le bord antérieur jusqu'au bout. Les ailes postérieures sont brunies avec l'extrémité plus pâle.

Pretoria.

Diffère des *Sc. apicalis*, Guér., *fasciatipennis*, Smith, *alaris*, Sauss., et *disparilis*, Kirby (peut-être toutes identiques) par ses poils blancs (qui ne le sont pas par vétusté). Néanmoins l'espèce pourrait être la même.

Discolia præstabilis, Saussure (sp. n.). (Tab. IV. fig. 9.)

Nigra. Antennæ, mandibulæ, sinus oculorum, pedes (coxis exceptis), abdominis segmenta 4^m–6^m, necnon 2^i et 3^i macula utrinque, rufa et rufo-hirta. Caput et thorax pilis a nigro ad rufum ludentibus hirsuta, fronte rufo-pilosella.

Abdominis segmenta 2^m et 3^m apice breviter, 3^m–5^m longe et abunde rufo-aureo fimbriata; 6^m appresso-pilosum. Alæ nigræ, cæruleæ; areola radiali secundam cubitalem haud superante, apice subperpendiculariter incurva. Tegulæ nigræ vel margine rufescente.

♀. Long. 18 mill., al. 12.

Tête petite, lisse, ponctuée en devant. Chaperon non strié, ayant sa protubérance médiane arrondie, lisse; son bord arqué réfléchi, point aplati. Poils du thorax paraissent noirs ou roux-dorés, suivant le jour sous lequel on les regarde; ceux du métathorax paraissent noirs. Thorax criblé en avant. Métanotum criblé de fortes ponctuations; métapleures un peu moins grossièrement ponctuées; la plaque postérieure concave, lisse, ponctuée; son arête supérieure subangulaire et presque vive au lobe médian; ses angles latéraux mousses; ses arêtes latérales prononcées mais émoussées. Abdomen ponctué; les poils de sa base noirs; les franges des 2e et 3e segments simples, courtes, partant d'un sillon placé très près du bord apical. La cellule radiale ayant son extrémité peu étroite, arquée, à peine oblique.

Pretoria.

Genus *Elis*, Fabr., Sauss.

Elis barbata, Saussure (apud Grandidier, Madagascar, Hyménoptères, p. 217, 1890).

Nigra; capite et thorace longe rufo-hirsutis; metathorace supra confertim punctulato, medio margine angulatim retroproducto, faciei posticæ marginibus haud acutis; abdominis segmentis 1°–4° albido-fimbriatis, 5° et 6° fusco-setosis; pedibus albido-pilosis; alis subhyalinis, venis ferrugineis, dimidia parte apicali nebulosa, subcyanescente.

♀. Long. 16 mill., al. 12.

Noire. Antennes et mandibules noires. Tête et thorax hérissés de longs poils d'un roux vif. Le mésonotum assez grossièrement ponctué. Écailles passant au testacé-roux. Métathorax en dessus densément et assez finement ponctué; sa plaque postérieure parsemée de petites ponctuations, et ayant toutes ses arêtes arrondies, sauf le milieu de l'arête supérieure, laquelle est tranchante et surplombe faiblement en formant un angle obtus.

Abdomen noir ou légèrement irisé, à reflets moirés, parsemé de poils blanchâtres, hérissé à sa base. Segments 2e et 3e semés de ponctuations très espacées ; 2e–5e offrant une zone marginale densément ponctuée et piligère, le devenant toujours plus fortement ; aux 2e–4e cette zone précédée par une zone médiane lisse et glabre, comme subcarénée transversalement aux 3e et 4e ; la base de ces segments faiblement ponctuée ; leurs côtés fortement criblés. En dessous, les segments 2e–4e ponctués à leur base et avec une ligne arquée de ponctuations piligères placée après le milieu, formant une sorte de sillon. Segments 1er–4e offrant, tant en dessus qu'en dessous, une frange de poils blanchâtres couchés, formant des bandes blanchâtres très nettes ; segments suivants garnis de soies brunes.

Pattes hérissées de poils blanchâtres ; les épines des tibias et des tarses d'un blanc roussâtre. Éperons des tibias postérieurs blanchâtres ; l'interne très long, fortement spatuliforme ; la spatule s'atténuant graduellement vers la base ; l'externe court, styliforme, cannelé, non dilaté.

Ailes légèrement enfumées, avec un faible reflet rose-violacé ; les nervures et leur entourage ferrugineux, ainsi que la côte jusqu'au bout de la cellule radiale ; celli-ci très obtuse, tronquée-arrondie, dépassant fort peu la 2e cubitale ; son bord apical formant, avec la côte, presque un angle droit. Le stigma nul.—*Var.* Les poils du thorax cendrés en dessous et sur les côtés.

Pretoria.

Espèce plus petite que l'*E. capensis*, intermédiare pour la livrée entre cette dernière et l'*E. rufa*, la tête et le thorax étant revêtus de longs poils roux comme chez cette dernière, l'abdomen orné de quatre bandes de poils blanchâtres comme chez la première.—Le grand éperon des tibias postérieurs est plus dilaté que chez ces espèces. Les nervures de l'aile sont comme chez l'*E. capensis*.

Genus *Mesa* *, Saussure.

Mesa, H. de Saussure, apud Grandidier, Madagascar, Hyménoptères, p. 244.

Mesa diapherogamia, Saussure (sp. n.). (Tab. IV. fig. 8.)
Nigra, nitida, cinereo-hirta; capite, antennarum articulis 1°, 2°, pronoto, tegulis, mesopleuris, mesonoti macula utrinque, scutello pedibusque, rufis ; alis fusco-violaceis, posticis basi breviter vitreis. Thorax supra sparse crasse punctatus ; mesopleuræ crassissime cribrosæ; metanotum subconvexum, læviusculum, impunctatum, per sulcum longitudinalem divisum, vel potius minute bisulcatum et tricarinulatum ; facie postica plana, leviter obliqua, obsolete punctata, canthis lateralibus acutis, margine supero-apicali valde rotundato, tota crasse, haud profunde punctata. Abdomen sparse punctatum; pygidium in longitudinem strigatum, apice obtusangulatum, angulo rotundato. Alarum venulatio ut in *M. atopogamia*, Sauss.
♀. Long. 17 mill., al. 12·5.

La tête est polie, non ponctuée en dessus, avec une tache noire sur les ocelles et l'espace sous les antennes noirâtre. Le pronotum n'est guère ponctué au milieu, mais il l'est fortement le long du bord postérieur, ce bord restant toutefois étroitement lisse, dépourvu de ponctuations, de même que le bord antérieur du mésonotum. Le mésonotum offre à côté de chaque écaille une tache longitudinale rousse.

Espèce très voisine de la *M. atopogamia* (Sauss. l. c.) de Zanzibar. Pretoria.

Familia MUTILLIDÆ.

Mutilla albistyla, Saussure (sp. n.). (Tab. IV. fig. 7.)
Tota profunde atra, atro-pilosa. Caput reticulato-punctatum, infra antennas profunde depressum. Antennæ crassæ, scapo punctato, anterius carinato. Thorax elongatus, ubique crassiuscule punctatus ; a latere anterius haud insigniter con-

---

* This genus has been established to include the *Plesiæ* of the Old World; these are distinguished principally by having the radial cell of the wing separate from the *costa* only at its end, while in the true *Plesiæ* (all American) this cell is separated from the costa in its whole length.

vexus. Mesonotum carinulatum, profunde grosse bisulcatum. Tegulæ dense punctatæ. Scutellum valde prominulum, rotundatum, reticulato-punctatum, a mesonoto profunde sejunctum; postice et utrinque per sulcum latum canaliculatum a metanoto separatum. Hoc in plano inferiore jacens, rotundatum, crasse reticulatum, per carinam longitudinalem divisum. Abdomen supra minus fortiter punctatum, nitidum, nigro-pilosum, haud velutinum; primum segmentum sat breve, depressum, trapezinum; ejus carina ventralis parum prominula, anterius dentem trigonalem efficiens, retro evanescens; secundum segmentum utrinque macula velutina transversa alba; segm. 6^m apice et 7^m albo-pilosa; hoc apice truncatum. Pedes nigro-pilosi, femoribus partim cinereo-pilosis; calcaribus tibiarum intermediarum et posticarum albidis. Alæ nigerrimæ, cærulcæ, areolis cubitalibus 3, venis recurrentibus 2; areola radiali apice haud truncata; 2^a cubitali quam tertia latiore; hæc paulo altior quam latior.

♂. Long. 20 mill., al. 19.
Pretoria.

Familia TENTHREDINIDÆ.

Genus *Athalia*, Leach.

Athalia bicolor, Saussure, sp. n.

Nigra; mesonoto pubescente; abdomine pedibusque flavo-testaceis vel fere aurantiis; tibiis tarsorumque articulis apice nigris; alis subvitreis, dimidia parte basali venis flavidis, dimidia parte apicali nebulosa, venis brunneis; venis costalibus stigmateque fuscis.
♀ ♂. Long. 8 mill., al. 8.

Antennes noires, souvent rousses en dessous. Tête et thorax noirs, revêtus d'un duvet soyeux gris. La tête en devant fortement pubescente, presque argentée. Chaperon, labre et mandibules blancs ou jaunâtres; celles-ci avec leur extrémité noire ou rousse et noire. Mésonotum pubescent, garni de poils brunâtres. Prothorax et métathorax en dessous plus ou moins jaune-testacés. Postécusson passant souvent au testacé au milieu et taché de noir; ses deux dents du bord antérieur blanches.—Abdomen jaune; l'espace lisse du 1er segment bien accusé; le fourreau de la scie noir.—Pattes avec

leurs hanches jaunes ; tibias et tarses annelés de noir au bout
des articles (par variété les pattes antérieures souvent non
annelées).

Pretoria, 1 ♀, 2 ♂.

Familia ICHNEUMONIDÆ.

Tribus PIMPLINÆ.

Genus *Hemipimpla*, Sauss.

Hemipimpla, Saussure, ap. Grandidier, Madagascar, Hymén. tab. 13. fig. 4.

Tête comprimée à vertex transversal ; sa face postérieure
perpendiculaire. La fossette frontale prononcée ; le vertex en
dos d'âne arrondi portant les ocelles rangés en triangle large et
très rapprochés.—*Abdomen* déprimé, ♀ subcomprimé à l'extrémité ; le dernier segment ventral fendu, emboitant la tarrière ;
celle-ci peu recouverte à sa base. (*Pimploïdæ*, Först.)—*Ailes* :
l'aréole grande, quadrangulaire, en rectangle oblique, pétiolée.
—*Pattes* grêles ; les fémurs non fusiformes. La base de la
tarrière un peu recouverte par le 6e segment ventral.

Thorax poli ; métathorax ponctué. Abdomen chagriné,
bosselé, les segments étranglés en gouttière après le milieu.
L'aréole de l'aile ayant son bord antérieur brisé avant le milieu,
son bord postérieur après le milieu.

Hemipimpla caffra, Saussure, sp. n.

Rufa ; antennis nigris, scapo subtus et imo apice ferrugineis ;
 alis ferrugineis, limbo apicali anguste irregulariter nigro ;
 posticis apice latius nigris, limbo postico angustissime nigro ;
 vagina nigra.

♀. Long. 14 mill., al. 13, oviposit. 4–5.

Tête passant au jaune, surtout en devant. Le front avec
deux fossettes. Le vertex arrondi, non bordé. Thorax poli.
Métathorax aplati-arrondi en arrière, et poli, en dessus lisse au
milieu, ponctué de chaque côté et sur les flancs.—Abdomen
fortement, densément ponctué-chagriné ; ses segments assez
fortement étranglés après le milieu. Tarses postérieurs bruns

en dessus.—L'aréole de l'aile rectangulaire ; la veine récurrente un peu arquée. Le bord postérieur de l'aile postérieure étroitement bordé de noir dans sa seconde moitié.

Pretoria, 1 ♀.

Hemipimpla calliptera, Saussure, sp. n.

Atra; capite sulfureo, fronte nigra; prothorace, mesopleuris antice, mesonoti marginibus lateralibus, testaceis; tegulis pedibusque anticis succincis, his brunneo-umbratis; alis aurantiis, binis apice late nigris.

♀. Long. 13·5 mill., al. 12, oviposit. 4.

Noir. Antennes noires. Le scape en dessous brunâtre, surtout à son bord apical. La tête en devant jaune-pâle, ainsi que la bouche, noire à sa face postérieure ; la face sous les antennes un peu brunie en milieu ; le front, avec les fossettes antennaires et le milieu du vertex, bruns. La fossette frontale transversale. L'occiput formant une arête de chaque côté des ocelles. Angles postérieurs du pronotum, une tache sur la partie antérieure des mésopleures et souvent une ligne horizontale vers le bas, ainsi que les bords latéraux du mésonotum jusqu'aux ailes, testacés. Écailles rousses. Métathorax poli et aplati en arrière, du reste criblé de fortes ponctuations, sauf sur la ligne médiane. —Abdomen fortement chagriné, densément couvert de ponctuations presque finement-réticuleuses.—Pattes des deux premières paires jaune-d'ambre, ombrées de brun en milieu des articles ; hanches noires ; les antérieures tachées de jaune en devant. Pattes postérieures noires ; tibias à leur base brièvement jaunes ; leurs éperons jaunes. Dernier article du tarse roux.— Ailes ferrugineuses ou orangées, avec une petite tache noire au bout du radius avant le stigma, et le bout de l'organe au-delà du stigma, noir ; la partie noire coupée perpendiculairement à son bord interne ; aux ailes postérieures la partie noire coupée obliquement à son bord interne. L'aréole un peu plus grande que chez la *H. caffra*; en rectangle un peu élargi, vers sa base, c.-à-d. vers le bord interne ; celui-ci un peu plus oblique que l'externe. La veine récurrente plus courbée.

Pretoria, 1 ♀.

INSECTA.

Tribus C r y p t i n æ.

Genus *Distantella*, gen. n.

Corpus gracile. Antennæ filiformes, ♀ haud nodosæ.—Caput elongato-trigonale, infra oculos sat longe productum. Mandibulæ arcuatæ, acutæ, haud dentatæ. Maxillæ in rostrum breve productæ. Palpi labiales 4-articulati; maxillares 5-articulati, articulis 3 ultimis brevioribus.

Occiput rotundatum. Ocelli magni, in trigonum latum exserti. Oculi elliptici, convexi, subsinuati. Frons supra antennas utrinque foveolata, margine laterali ad sinum oculorum reflexo. Facies infra antennas quam frons altior, in medio parallele prominula, superne subtuberculata. Clypeus elevatus, apice totus truncatus.

Thorax compressus, rugatus, superne punctatus, antice attenuatus. Mesonotum longius quam latius, bisulcatum. Scutellum convexum. Metathorax brevis, rotundatus, rugosus, postice declivis, basi transverse angulato-carinulatus; stigmata rimata.

Abdomen compressum, longe petiolatum. Primum segmentum filiforme, thorace paulo brevius, arcuatum, subdepressum; ejus stigmata tuberculata, paulum pone medium aperta; petiolus pone illa gradatim leviter dilatatus, sed nullomodo campanulatus. Segmentum secundum elongato-infundibuliforme, quam latius duplo longius, haud compressum. Abdomen reliquum compressum. Segmentum 7^m subtus apertum, 8^m subtus fissum, supra apice truncatum. Terebra abdominis longitudine.

Pedes antici et intermedii graciles; tibiæ anticæ crassiusculæ, superne dense spinulosæ; intermediæ graciles, supra et subtus spinulis minimis seriatim instructæ. Pedes postici longi; femora basin versus attenuata; tibiæ crassæ, compressæ, superne et subtus rotundato-carinatæ, basi attenuatæ, latere externo plano, bisulcato, latere interno convexo. Tarsi subtus spinulosi; metatarsi paralleli, reliquos articulos computatos fere æquantes.

Alæ coloratæ. Areola pentagonalis, sat grandis, quam latior valde altior, antice paulum coarctata, margine postico angulato. Stigma angustum, acute lanceolatum. Margo internus areolæ discoïdalis posticæ vix ultra angulum basalem areolæ cubito-discoïdalis exsertus; hic angulus acutus, rotundatus.— In alis posticis venula transverso-discoïdalis obtusangulatim fracta, venam discoïdalem ante medium emittens.

Je ne saurais indiquer les analogies de ce genre avec d'autres.

Je ne crois pas pouvoir le placer ailleurs que dans la tribu des Cryptiens.

Distantella trinotata, Saussure, sp. n.

Nigra; antennis subtus rufis, ultra medium annulo longo aurantio ; facie infra antennas clypeoque rufis ; pedibus anticis fuscescentibus, tibiis subtus rufidis ; posticis flavis, apice nigro, tarso flavo, metatarso nigro, apice flavo ; 5° articulo nigro; abdominis segmentis 5°-8° aurantiis ; terebra nigra ; alis aterrimis, violascentibus, punctis hyalinis aliquibus, ornatis.

♀. Long. 25 mill., al. 20, petiol. 5·6, terebr. 15, antenn. 15, tib. post. 8.

Tête finement et densément ponctuée. La face revêtue d'un fin duvet soyeux gris, faiblement cannelée en longueur de chaque côté ; son milieu formant une bande un peu saillante, parallèle, occupant toute sa longueur et portant au sommet un petit tubercule allongé. Chaperon en forme de cloche, aussi haut que large en bas, aussi long que la face, finement pointillé au sommet, strié au bas d'une manière un peu convergente ; le bord inférieur transversal, finement lisse. Palpes bruns ; les derniers articles roussâtres, l'avant-dernier le plus court. Méso-notum distinctement ponctué, assez densément, étroit en avant ; ses sillons non prolongés en arrière. Écusson poli, semé de ponctuations éparses. Flancs rugueux, densément rugulés de strioles sinueuses. Métapleures rugulés et ponctués ; méta-notum rugueusement ponctué, le devenant grossièrement et réticuleusement à son extrémité postérieure. La carène trans-verse oblique de chaque côté, formant au milieu un angle obtus arrondi dirigé en avant, contigu au sillon postscutellaire. Stigmates en boutonnière.

Pétiole lisse, pointillé ; ses tubercules très prononcés, arrondis.

Abdomen très finement pointillé-striolé. Stigmates du 2ᵉ segment placés un peu en arrière du milieu ; ceux du 3ᵉ placés avant le milieu ; ceux du 5ᵉ tout à sa base. Sixième segment ventral transversal à bord postérieur peu arqué.—Pattes finement pointillées. Tibias postérieurs semés à leur face ex-terne d'épines microscopiques brunes. Éperons noirs, à pointe rousse ; ceux des tibias antérieurs roux. Griffes noires.

Ailes noires, avec de petits espaces hyalins tout à leur base et ornées de plusieurs petits points hyalins. Le bord postérieur de l'aréole ayant son angle placé en son milieu. La veine récurrente presque droite. Les points hyalins placés, aux ailes antérieurs : *un* sur l'extrémité postérieure de chacune des deux premières veines transverses, au contact de la veine anale, et 3 autour de l'aréole ; l'une sur l'angle postérieur-externe de l'aréole ; les deux autres plus grands, allongés, sur la veine cubitale et la veine récurrente, à quelque distance de l'aréole ; ces 3 taches toutes à cheval sur les nervures, qui restent noires et les partagent.—Aux ailes postérieures un point sur l'extrémité postérieure de la veinule transverso-cubitale et de la veinule transverso-discoïdale.

Pretoria, 1 ♀.

LEPIDOPTERA.

Rhopalocera.

I found no undescribed Butterflies around Pretoria, nor did I very much expect to do so, for when a specialist like Mr. Roland Trimen has presided for years at the Cape Town Museum, the captures made in South Africa have naturally found their way to him, and it would be well if such still continued to be the practice, at least for the present, and thus much synonymy might be prevented. These remarks apply only to the Butterflies, and not to other groups or orders of the South-African Insecta, which can only thoroughly be worked out in Europe.

On the open plains which surround Pretoria butterfly-life is not seen in much profusion, nor are many species to be found. In the Danainæ *Danais chrysippus* flies all the year round, throughout the warm, wet, and cool dry seasons ; both its varietal forms may also here be obtained, but they are very scarce, and constitute no appreciable proportion to the dominant form of the species (see *ante*, p. 65). In the Satyrinæ I only met with the genus *Ypthima* in the warmer lowlands of Zoutpansberg, and did not see a single example about Pretoria. Among the Nymphalinæ that world-wide (excluding South America, New Zealand, and Australia) distributed butterfly

Pyrameis cardui is found on the wing throughout the year, and its constancy of appearance seems also equalled by *Junonia cebrene*. The protective resemblance of *Hamanumida dædalus* has been described (*ante*, p. 41) ; but the lepidopterist has still much to observe on this point with other species, for in a world governed by unvarying natural laws, causation is now seen to have a new meaning, and not a butterfly possesses its peculiar coloration outside the rule of natural selection. Our admiration expressed for beauty alone is simply a confession of our ignorance.

In the Pierinæ the genus *Teracolus* frequents the warmer valleys of Natal and Zoutpansberg, and is moderately scarce around Pretoria. *Colias electra* is apparently on the wing throughout the year, and can be found alike in gardens and across the bare veld.

Some of the butterflies are exceedingly scarce and local, and it was only on one restricted spot near Pretoria that I ever saw the beautiful lycænid *Iolaus bowkeri*, but in this narrow strip of scrub it could always be found during the season. I have appended the dates at which I found the species, but of course these in most cases simply relate to my own experience, and do not define the total period of the insect's appearance.

Fam. NYMPHALIDÆ.

Subfam. DANAINÆ.

Danais chrysippus, Linn. (Throughout the year) Pretoria.
   Var. *alcippus*, Cram. Pretoria.
   Var. *dorippus*, Klug. Pretoria.

Subfam. SATYRINÆ.

Ypthima asterope, Klug. (May) Zoutpansberg.
Pseudonympha narycia, Wallengr. (Dec.) Pretoria.
Pseudonympha vigilans, Trim. (Dec.) Pretoria.
Mycalesis perspicua, Trim. (May) Zoutpansberg.
Melanitis leda, Linn. (Aug. & Sept.) Pretoria.

Subfam. ACRÆINÆ.

Acræa horta, Linn. (Nov.) Pretoria.
Acræa neobule, Doubl. (Feb. & Sept.) Pretoria,
   (May) Zoutpansberg.

Acræa violarum, Boisd. (May) Zoutpansberg.
Acræa nohara, Boisd. (Jan.) *Durban, Natal.*
Acræa doubledayi, Guér. (May) Zoutpansberg.
Acræa caldarena, Hewits. (May) Zoutpansberg.
Acræa natalica, Boisd. (Jan.) *Durban, Natal.*
Acræa anemosa, Hewits. (Sept.) Pretoria.
Acræa encedon, Linn. (May) Zoutpansberg.
Acræa rahira, Boisd. (Feb.) Pretoria.
Acræa buxtoni, Butl. (Jan. to May) Zoutpansberg, Pretoria, and *Natal.*

Planema esebria, Hewits. (Jan.) *Durban, Natal.*

Subfam. NYMPHALINÆ.

Atella columbina, Cram. (Mar.) Pretoria.
Pyrameis cardui, Linn. (Throughout the year) Pretoria.
Junonia cebrene, Trim. (Throughout the year) Pretoria.
Junonia clelia, Cram. (May) Zoutpansberg.
Junonia böopis, Trim. (Feb.) Pretoria.
Precis cloantha, Cram. (Aug. & Mar.) Pretoria.
Precis ceryne, Boisd. (May) Zoutpansberg.
Precis sesamus, Trim. (Feb.) Pretoria.
Precis archesia, Cram. (Feb. & Mar.) Pretoria.
Precis natalica, Feld. (May) Zoutpansberg,
  (Jan.) *Durban, Natal.*
Salamis anacardii, Linn. (Jan.) *Durban, Natal.*
Eurytela hiarbas, Dru. (Jan.) *Durban, Natal.*
Hypanis ilithyia, Dru. (May) Zoutpansberg,
  (Sept. & Mar.) Pretoria.
Neptis agatha, Cram. (May) Zoutpansberg.
Hypolimnas misippus, Linn. (Jan. & Mar.) Pretoria,
  (Jan.) *Durban, Natal.*
Hamanumida dædalus, Fabr. (Aug. to June) Pretoria,
  (May) Zoutpansberg.

Fam. LYCÆNIDÆ.

Catochrysops osiris, Hopff. (May) Zoutpansberg.
Tarucus telicanus, Lang. (May) Zoutpansberg,
  (Oct. to Mar.) Pretoria.

Castalius melæna, Trim.   (May) Zoutpansberg.
Azanus jesous, Guér.   (Oct.) Pretoria.
Cigaritis leroma, Wallengr.   (Oct.) Pretoria.
Lycænesthes liodes, Hewits.   (Feb.) Pretoria.
Iolaus bowkeri, Trim.   (Feb.) Pretoria.
Chrysorychia amanga, Westw.   (Mar.) Pretoria.
Zeritis orthrus, Trim.   (Sept.) Pretoria.
Pentila tropicalis, Boisd.   (Jan.) *Durban, Natal.*
Alæna amazoula, Boisd.   (Jan.) *Durban, Natal.*

Fam. PAPILIONIDÆ.

Subfam. PIERINÆ.

Terias brigitta, Cram.   (May) Zoutpansberg,
                                         (Aug., Sept., Oct.) Pretoria.
Terias zoë, Hopff.   (May) Zoutpansberg,
                                         (Dec., Jan., & Feb.) Pretoria.
Mylothris agathina, Cram.   (May) Zoutpansberg,
                                         (Feb. & Mar.) Pretoria.
Pieris mesentina, Cram.   (May) Zoutpansberg,
                                         (Aug. to June) Pretoria.
Pieris gidica, Godt.   (Jan.) *Durban, Natal.*
Pieris severina, Cram.   (May) Zoutpansberg.
Herpenia eriphia, Godt.   (Feb.) Pretoria.
Teracolus subfasciatus, Swains.   (May) Zoutpansberg.
Teracolus eris, Klug.   (Feb.) Pretoria.
Teracolus evenina, Wallengr.   (Feb.) Pretoria.
Teracolus achine, Cram.   (May) Zoutpansberg,
                                         (Nov.) Pretoria.
Teracolus gavisa, Wallengr.   (May) Zoutpansberg,
                                         (Jan.) *Durban, Natal.*
Teracolus omphale, Godt.   (May) Zoutpansberg,
                                         (Jan.) *Durban, Natal.*
Teracolus theogone, Boisd.   (May) Zoutpansberg.
Teracolus phlegetonia, Boisd.   (May) Zoutpansberg.
Teracolus vesta, Reiche.   (May) Zoutpansberg.
Colias electra, Linn. (Probably throughout the year) Pretoria.
Eronia leda, Boisd.   (Jan.) *Durban, Natal.*
Catopsilia florella.   (May) Zoutpansberg.

INSECTA. 235

Subfam. PAPILIONINÆ.

Papilio morania, Angas.   (Jan.) *Durban, Natal.*
Papilio demoleus, Linn.   (Sept.) Pretoria,
                                             (Jan.) *Durban, Natal.*

Fam. HESPERIDÆ.

Cyclopides willemi, Wallengr.   (Feb.) Pretoria.
Pyrgus vindex, Cram.   (Sept.) Pretoria.
Pyrgus diomus, Hopff.   (Aug.) Pretoria.
Pamphila hottentota, Latr.   (May) Zoutpansberg,
                                             (Oct.) Pretoria.
Hesperia forestan, Cram.   . (Feb.) Pretoria.

HETEROCERA.

In Sphingidæ one is reminded of home when catching *Protoparce convolvuli* or breeding *Acherontia atropos*, but the hawk-moths are not common, *Macroglossa trochilus* being the most abundant species. Among the day-flying Agaristidæ *Pais decora* frequents the bare veld, but in single examples; the genus *Xanthospilopteryx* I only found in the wooded districts of Pretoria. Another moth that frequents the open veld is *Deiopea pulchella*, remarks on whose habits have been previously made (*ante* p. 68). *Petovia* is a genus not often found, but its species have a slow conspicuous flight, and I took *P. dichroaria* flying in the busy streets of the town of Pretoria. Many moths come to light which are not seen otherwise, and, when making long coach-journeys, I have frequently boxed a good specimen at the lamp which illuminated the frugal meal obtained at the lone posting-houses on the bare and dreary veld. Even Saturniidæ come to light, as I found *Urota sinope* to do at Durban.

In the identification of the Heterocera I have received much assistance from Messrs. Butler, Kirby, and Warren at the British Museum; and among the Micro-Lepidoptera, I have had the great help of Lord Walsingham, and his Secretary, Mr. J. Hartley Durrant, has described a new genus and species.

Fam. SPHINGIDÆ.

Hemaris hylas, Linn.	(May) Zoutpansberg.
Macroglossa trochilus, Hübn.	(Feb.) Pretoria.
Chærocampa schenki, Mösch.	(Sept.) Pretoria.
Acherontia atropos, Linn.	(Mar.) Pretoria.
Protoparce convolvuli, Linn.	Pretoria.

Fam. ZYGÆNIDÆ.

Syntomis khulweinii, Lefèbv.	(Jan.) *Durban, Natal.*
Trianeura fulvescens, Walk.	(May) Zoutpansberg and Pretoria.
Anteris zelleri, Wallengr.	(May) Zoutpansberg.
Neurosymploca concinna, Dalm.	Pretoria.
Neurosymploca agria, sp. n.	Pretoria.
Arthileta cloeckneria, Cram.	Pretoria.
Euchromia africana, Butl.	(Jan.) *Durban, Natal.*

Description.

Neurosymploca agria, sp. n. (Tab. II. fig. 9.)

Anterior wings black, with six large ochraceous spots, arranged transversely and obliquely in pairs, two at base, two at centre, and two subapical, the last spot long and subquadrate; posterior wings carmine-red, apical and outer margin (not reaching anal angle) black, and a small discal black spot near centre of upper margin of cell. Wings beneath as above. Head, antennæ, and thorax black, a large ochraceous spot on each side of pronotum; abdomen carmine-red, its extreme apex fuscous; legs black, tinged beneath with ochraceous.

Exp. wings 29 millim. (*W. L. D.*)

Fam. AGARISTIDÆ.

Pais decora, Linn.	Pretoria.
Xanthospilopteryx superba, Butl.	(Feb.) Pretoria.

Fam. NYCTEMERIDÆ.

Leptosoma apicalis, Walk.	(Jan.) *Durban, Natal.*

Fam. LITHOSIIDÆ.

Lithosia ? fumeola, Walk.	(Feb.) Pretoria.
Siccia caffra, Walk.	Pretoria.
Petovia marginata, Walk.	Pretoria.
Petovia dichroaria, Herr.-Sch.	Pretoria.
Deiopeia pulchella, Linn.	Pretoria.

Fam. ARCTIIDÆ.

Teracotana submacula, Walk.	Pretoria.
Binna linea, Walk.	Pretoria.
Binna madagascariensis, Butl.	Pretoria.
Anace lateritia, Herr.-Sch.	(Jan.) Heidelberg.
Egybolia vaillantina, Stoll.	(Jan.) Durban, Natal.
Decimia bicolorata, Walk.	(Jan.) Wakkerstroom.

Fam. LIPARIDÆ.

Lælia adspersa, Herr.-Sch.	Pretoria.

Fam. NOTODONTIDÆ.

Calpe apicalis, Walk.	Pretoria.

Fam. LIMACODIDÆ.

Crothæma decorata, sp. n. (Tab. II. fig. 6.)

Anterior wings above violaceous; the basal costal margin and base, an irregularly shaped spot below discal centre, and an apical streak ochraceous; a much-waved transverse fascia commencing at about centre of costal margin and reaching discal spot, which it inwardly margins, and a large subapical patch creamy white, the last partly contains the apical ochraceous streak; outer margin with small violaceous spots and lunules; fringe ochraceous. Posterior wings and body warm bright ochraceous, wings and body beneath bright ochraceous, the anterior wings with a strongly reddish tinge.

Exp. wings 36 millim.

Allied to *C. sericea,* Butl., from Madagascar.   (*W. L. D.*)

Fam. LASIOCAMPIDÆ.

Lebeda aculeata, Walk.	(Aug.) Pretoria.

Fam. SATURNIIDÆ.

Gynanisa maia, Klug.	(Sept.) Pretoria.
Urota sinope, Westw.	(Jan.) *Durban, Natal.*
Aphelia apollinaris, Boisd.	(Jan.) *Durban, Natal.*

Fam. HEPIALIDÆ.

Gorgophis libania, Cram.	Pretoria.

NOCTUES.

Fam. LEUCANIIDÆ.

Leucania apparata, Wallengr.	Pretoria.
Leucania percussa, Butl.	Pretoria.
Leucania substituta, Wallengr.	Pretoria.
Leucania amens, Walk. (nec Guér.).	Pretoria.
Axylia cinctothorax, Walk.?	Pretoria.

Fam. ACONTIIDÆ.

Tarache caffraria, Cram.	Pretoria.

Fam. NOCTUIDÆ.

Agrotis biconica, Koll.	Pretoria.

Fam. APAMIIDÆ.

Mamestra breviuscula, Walk.	Pretoria.
Mamestra renisigna, Walk.	Pretoria.
Ozarba dubitans, Walk.	Pretoria.
Ozarba densa, Walk.	Pretoria.

Fam. PLUSIIDÆ.

Plusia acuta, Walk.	Pretoria.

Fam. POLYDESMIDÆ.

Polydesma landula, Guen.	(May) Zoutpansberg.

Fam. OMMATOPHORIDÆ.

Cyligramma latona, Cram.	Pretoria.

Fam. OPHIUSIDÆ.

Ophisma pretoriæ, sp. n.   Pretoria.
Ophisma croceipennis, Walk.   (May) Zoutpansberg.
Sphingomorpha monteironis, Butl.   Pretoria.

Description.

Ophisma pretoriæ, sp. n. (Tab. II. fig. 7.)

Anterior wings above dull brownish ochraceous with a glossy hue, crossed by a dark curved linear fascia at about one third from base, slightly directed outwardly; two much-waved and inwardly-curved linear fasciæ crossing wing beyond cell; basal, costal, and outer margins dusky brown; two small black spots at extremity of cell, and an outer submarginal series of black dots: posterior wings pale brownish, the outer margins fuscous. Wings beneath paler than above; anterior wings with an elongate spot at end of cell, and a curved waved fascia crossing wing between end of cell and apex; outer margin dull ochraceous, with darker elongate marks between the nervures: posterior wings with a dark spot in cell, two linear fasciæ crossing wing beyond cell, the innermost nearly even, the outermost much waved, hind margin as in anterior wing. Body and legs dull brownish ochraceous.

Exp. wings 53 millim.

This species appears to be most closely allied to *O. senior*, Walk. (*W. L. D.*)

DELTOIDES.

Fam. HERMINIIDÆ.

Epizeuxis æthiops, sp. n. (Tab. II. fig. 2.)

Anterior wings above ashy grey, mottled with fuscous; six distinct dark spots on costal margin, which are continued more or less distinctly across wing in waved fasciæ, before the apical fascia the colour is irregularly and narrowly warm ochraceous, a series of small dark spots on outer margin; fringe long and ashy grey: posterior wings above pale ashy grey, streaked on basal area with dark fuscous. Wings beneath pale ashy grey, with the dark markings of the upper side practically absent.

Body above concolorous with wings, beneath much paler in hue; anterior tibiæ dark fuscous.

Exp. wings 18 millim.

The most closely allied species to *E. æthiops* appears to be the *E. maculifera*, Butl., from Dharmsala. (*W. L. D.*)

GEOMETRITES.

Gonodela amandata, Walk.	Pretoria.
Nassunia bupaliata, Walk.	Pretoria.
Cidaria pudicata, Guen.	Pretoria.
Osteodes turbulentata, Guen.	Pretoria.
Sterrha sacraria, Linn.	Pretoria.
Eubolia deercana, Walk.	Pretoria.
Eubolia proxaulkaria, Walk.	Pretoria.
Eubolia punicaria, Guen., var.	Pretoria.
Lycauges donovani, sp. n.	Pretoria.
Acidalia gazella, Wallengr.	Pretoria.
Cænina pæcilaria, Herr.-Sch.	Zoutpansberg.

Description.

Lycauges donovani, sp. n. (Tab. II. fig. 4.)

Wings above pale ochraceous, dusted with fuscous; anterior wings with a fuscous fascia commencing at apex and terminating on inner margin near centre, a zigzag fuscous line crosses cell to inner margin, outer margin with a double series of small brown spots; posterior wings crossed by three transverse dark fasciæ, the central one very sinuate, outer margin with a double series of small brown spots. Wings beneath much paler in hue; cells of both wings with a small black spot at apices; basal areas darker, a dark oblique fascia crossing both wings, and a dark waved outer submarginal fascia; extreme outer margin minutely spotted with fuscous. Body above concolorous with wings, but abdomen banded with fuscous; body beneath concolorous with wings; legs dull ochraceous.

Exp. wings 25 millim.

I have named this species after T. Donovan, to whose exertions at Pretoria I am indebted for the acquisition of many small Heterocera. (*W. L. D.*)

Pyrales.

Fam. Botydidæ.

Acharana otreusalis, Walk.	Pretoria.
Paliga infuscalis, Zell.	Pretoria.
Pionea africalis, Guen.	Pretoria.

Fam. Margaronidæ.

Euclasta warreni, sp. n.	Pretoria.

Fam. Spilomelidæ.

Haritala quaternalis, Zell.	Pretoria.

Fam. Pyralidæ.

Pyralis farinalis, Linn.	Pretoria.
Aglossa basalis, Walk.	Pretoria.

Fam. Scoparidæ.

Nomophila noctuella, Schiff.	Pretoria.
Tritæa lacunalis, Zell.	Pretoria.

Description.

Euclasta warreni, sp. n. (Tab. II. fig. 5.)

Anterior wings above silvery grey, more or less shaded with brownish, excepting a long discal streak which is shining and immaculate; a narrow basal subcostal line, a large irregular fascia occupying upper half of cell and extending to apex, two spots—sometimes fused—on lower apical area, and some marks near inner margin dark fuscous. Posterior wings pale greyish hyaline, margins fuscous—apical broadly and posterior very narrowly. Wings beneath much paler in hue. Body above greyish brown, head with three longitudinal silvery lines in front of eyes, margins of thorax and upper surface of abdomen shaded with silvery grey.

Exp. wings 35 millim. (*W. L. D.*)

Fam. TINEIDÆ.

(By JNO. HARTLEY DURRANT, F.E.S., M.Soc.Ent.Fr.)

Subfam. TINEINÆ.

Sematocera, gen. nov.

($\sigma\hat{\eta}\mu a$ = a sign, $\kappa\acute{\epsilon}\rho a\varsigma$ = a horn.)

Type, ♂. *Sematocera fuliginipuncta.* (Tab. IV. figs. a,b,c,d,e.*)

Antennæ slightly more than half the length of the fore wings; bipectinate (5), each pectination biciliate.

Maxillary palpi folded (concealed in the clothing of the labial palpi).

Labial palpi, second joint recurved, densely clothed beneath, almost concealing the porrect apical joint.

Haustellum not apparent.

Head somewhat roughly clothed above.

Fore wings, costa slightly arched, apex rounded, apical margin oblique. *Neuration*: 12 veins, all separate; 7 to apex, continued through the cell to between 10 and 11, and forming a supplementary cell, from which arises 8, 9, and 10; another internal vein from the base is forked near the end of the cell, one branch going to between 5 and 6, the other to between 3 and 4; 1 furcate at base.

Hind wings slightly wider than the fore wings, subovate. *Neuration*: 8 veins, all separate; the internal vein is forked, one branch going to between 3 and 4, the other to between 6 and 7; 1 *b* furcate at base.

Abdomen moderate; anal appendages strongly developed.

Legs, hind tibiæ clothed on their inner side.

This genus differs from *Scardia*, Tr. (to which it is nearly allied), in the form of the palpi and the bipectinate antennæ. *Euplocamus*, Latr., which has similar antennæ, is distinguished by the structure of the palpi and by having veins 7 and 8 of the fore wings from a common stem.

The possession of folded maxillary palpi separates it from *Lasioctena*, Meyr. (J. H. D.)

* 4 *a*. Neuration; 4 *b*. Head; 4 *c*. Inner side of labial palpus, showing the concealed maxillary palpi; 4 *d*. Part of antennæ; 4 *e*. Three joints of antennæ, highly magnified.

Sematocera fuliginipuncta, sp. n.   (Tab. IV. fig. 4.)

Antennæ dark fuscous.
Palpi brownish ochreous.
Head ochreous.
Thorax dull white, with some blackish scales at the base and apex of the tegulæ, and a few before its junction with the abdomen.
Fore wings dull white, with soot-black spots which tend to form irregular fasciæ; at one third from the base a reduplicated fascia of black spots arises from the costa and, inclining obliquely outwards to the outer edge of the cell, is angulated downward to the dorsal margin; the costa is spotted with black from the base to the origin of these fasciæ, and near the base of the wing is another outwardly angulated fascia of spots; the apical third of the wing is irrorated with black spots, which, however, also arrange themselves in fasciaform lines; cilia dull white, with some fuscous scales intermixed. The markings can be distinctly traced on the underside.
Hind wings pale greyish ochreous, with cilia of the same colour.
Abdomen pale greyish ochreous.
Hind legs pale ochreous; tarsi unspotted.
Exp. al. 22 millim.
Hab. Transvaal: Pretoria (*Distant*).
Type ♂ .   (J. H. D.)

Tinea, L.

Tinea vastella, Z.

Transvaal: Pretoria (*Distant*).
A single female in poor condition.

Tinea, sp.

Transvaal: Pretoria (*Distant*).
A single specimen without abdomen is in poor condition. It is smaller than any specimen of *vastella* that I have yet seen; but if not that species it is, at least, so closely allied to it that a comparative description would be useless. The head is bright

canary-yellow, and the underside of the fore wings deep purplish fuscous, with the apical third of the costa and the cilia dull ochreous. Exp. al. 18 millim. (*J. H. D.*)

DIPTERA.

(By ERNEST E. AUSTEN, Zool. Dept. Brit. Mus.)

Fam. TABANIDÆ.

Pangonia subfascia, Wlk.
Silvius denticornis, Wied.
Tabanus socius, Wlk.
Atylotus, sp. (allied to *Tabanus* (*Atylotus*) *diurnus*, Wlk.).

Fam. ASILIDÆ.

Microstylum dispar, Lw.
Microstylum, sp. (? nov.—Closely allied to *M. gulosum*, Lw.).
Lophonotus, sp. (? nov.).
Lophonotus, sp. (? nov.—Allied to *L. ustulatus*, Lw.).

Fam. MUSCIDÆ.

Calliphora marginalis, Wied.
Musca domestica, L.

Fam. SARCOPHAGIDÆ.

Sarcophaga, sp.

Fam. HIPPOBOSCIDÆ.

Hippobosca rufipes, Olfers.

[This species not only attacks horses, but also frequently attached itself to my neck. (*W. L. D.*)]

RHYNCHOTA.

HETEROPTERA.

The open plains which surround Pretoria are not calculated to prove a home for many species of this order, though the tall blooming grasses and Asclepiads (*Gomphocarpus*) are particularly attractive to Lygæids (the rare *Lygæus septus*, Germ., was thus found), whilst many species, especially Reduviids, were only met with under stones and pieces of quartzite. These are favourite situations during the dry season for many insects, and even Pentatomidæ are no exception to the rule, but in a bare and treeless region find their only shelter under the rocky débris which strew the plains. On the wing, species of the genus *Aspongopus* and *Anoplocnemis curvipes* are the most abundant, and appearing early, fly throughout the summer season, whilst during the same period the stridulation of *Platypleura divisa* is heard from most of the willow trees that abound in Pretoria. I was surprised to find how many small beetles become the prey of the *Reduviids*, and the rostrum of *Physorhynchus patricius* produces more intense pain than the bite or puncture of any other insect with which I am acquainted.

Small as the material collected is, I am able to add a new genus and fifteen species hitherto unrecorded in entomological literature.

Fam. PENTATOMIDÆ.

Subfam. SCUTELLERINÆ.

Steganocerus multipunctatus, Thunb.	Pretoria.
Callidea natalensis, Stål.	Pretoria.

Subfam. CYDNINÆ.

Dismegistus fimbriatus, Thunb.	Waterberg.

Subfam. ASOPINÆ.

Glypsus mæstus, Germ.	Pretoria.
Glypsus conspicuus, Hope.	Pretoria.
Cimex figuratus, Germ., var. n.	Pretoria.

Subfam. PENTATOMINÆ.

Cænomorpha nervosa, Dall.	Pretoria.
Paramecocoris ventralis, Germ.	Pretoria.
Paramecocoris atomarius, Dall.	Pretoria.
Tropicorypha corticina, Germ.	Pretoria.
Holcostethus obscuratus, sp. n.	Pretoria.
Halyomorpha capitata, sp. n.	Zoutpansberg.

Halyomorpha pretoriæ, sp. nov.   Pretoria.
Veterna sanguineirostris, Thunb.   Pretoria.
Veterna pugionata, Stål.   Pretoria.
Veterna patula, sp. n.   Pretoria.
Veterna subrufa, Stål.   Pretoria.
Caura rufiventris, Germ.   Pretoria.
Dichelocephala lanceolata, Fabr.   Pretoria.
Carbula trisignata, Germ.   Pretoria.
Agonoscelis versicolor, Fabr.   Pretoria.
Agonoscelis erosa, Hope.   Pretoria.
Agonoscelis puberula, Stål.   Pretoria.
Bagrada hilaris, Burm.   Pretoria.
Nezara viridula, Linn.   Pretoria.
Nezara capicola, Hope.   Pretoria.
Antestia transvaalia, sp. n.   Pretoria.
Piezodorus purus, Stål.   Pretoria.

Subfam. DINIDORINÆ.

Aspongopus japetus, Dist.   Waterberg and Pretoria.
Aspongopus nubilus, Hope.   Pretoria.

Subfam. PHYLLOCEPHALINÆ.

Gellia angulicollis, Stål.   Pretoria.

Fam. COREIDÆ.

Subfam. COREINÆ.

Elasmopoda undata, Dall.   Durban, Natal.
Hoploterna valga, Linn.   Pretoria.
Anoplocnemis curvipes, Fabr.   Pretoria and *Durban*.
Homœocerus magnicornis, Burm.   Pretoria.
Homœocerus annulatus, Thunb.   Pretoria.
Rhyticoris terminalis, Burm.   Durban, Natal.
Plinachtus pungens, Thunb.   Pretoria.
Plinachtus falcatus, sp. n.   Pretoria.
Cletus ochraceus, Herr.-Schäff.   Pretoria.
Acanthocoris lugens, Stål.   Pretoria.

Subfam. ALYDINÆ.

Mirperus jaculus, Thunb.   Pretoria.

INSECTA.

Fam. LYGÆIDÆ.

Subfam. LYGÆINÆ.

Oncopeltus famelicus, Fabr.	Pretoria.
Lygæus elegans, Wolff.	Pretoria.
Lygæus trilineatus, Fabr.	Pretoria.
Lygæus planitiæ, sp. n.	Pretoria.
Lygæus desertus, sp. n.	Pretoria.
Lygæus rivularis, Germ.	Pretoria.
Lygæus campestris, sp. n.	Pretoria.
Lygæus septus, Germ.	Pretoria.
Aspilocoryphus fasciativentris, Stål.	Pretoria.
Lygæosoma villosula, Stål.	Pretoria.
Transvaalia lugens, gen. et sp. n.	Pretoria.
Nysius novitius, sp. n.	Pretoria.
Pamera proxima, Dall.	Pretoria.
Pachymerus apicalis, Dall.	Pretoria.
Dieuches armipes, Fabr.	Pretoria.
Dieuches patruelis, Stål.	Pretoria.

Fam. PYRRHOCORIDÆ.

Subfam. PYRRHOCORINÆ.

Dermantinus limbifer, Stål.	Zoutpansberg and Pretoria.
Dermantinus lugens, Stål.	Pretoria.
Scantius forsteri, Fabr.	Pretoria.

Fam. REDUVIIDÆ.

Subfam. REDUVIINÆ.

Pantoleistes princeps, Stål.	Pretoria.
Reduvius erythrocnemis, Germ.	Pretoria.
Reduvius pulvisculatus, sp. n.	Pretoria.
Reduvius sertus, sp. n.	Waterberg and Pretoria.
Reduvius capitalis, sp. n.	Pretoria.
Reduvius rapax, Stål.	Pretoria.
Physorhynchus crux, Thunb.	Pretoria.
Physorhynchus patricius, Stål.	Pretoria.

Subfam. PIRATINÆ.

Pirates lugubris, Stål.	Pretoria.
Pirates conspurcatus, sp. n.	Pretoria.

Subfam. ACANTHASPIDINÆ.

Edocla quadrisignata, Stål.	Pretoria.

HOMOPTERA.

Fam. CICADIDÆ.

Platypleura divisa, Germ.	Pretoria.
Platypleura punctigera, Walk.	*Durban, Natal.*
Tibicen carinatus, Thunb.	Pretoria.
Tibicen undulatus, Thunb.	Pretoria.

Fam. CERCOPIDÆ.

Subfam. CERCOPINÆ.

Locris transversa, Thunb.	*Durban, Natal.*
Locris arithmetica, Walk.	Pretoria.

Subfam. APHROPHORINÆ.

Poophilus actuosus, Stål.	Pretoria.

Notes and Descriptions.

Cimex figuratus, Germ., var. (Tab. III. fig. 1.)

Asopus figuratus, Germ. in Silb. Rev. Ent. v. p. 185, no. 132 (1837).

Bright blue, shining, corium paler and more greenish in hue; head with the central lobe obscurely marked with luteous; pronotum with the anterior and lateral margins and some scattered spots on anterior half of disk luteous; corium with base of lateral margin luteous; scutellum with a small central basal spot and the apical margin very narrowly luteous; connexivum luteous, spotted with dark indigo-blue; membrane brassy brown, with its apex hyaline. Body beneath bright shining purplish blue; lateral margins of sternum, a spot on each side of metasternum, rostrum, coxal spots, bases of femora, apex of spine on anterior femora, a broad central annulation to posterior tibiæ, central transverse fasciæ and marginal spots to abdomen, luteous. Antennæ and apex of rostrum blackish.

Long. 13 millim. (*W. L. D.*)

Holcostethus obscuratus, sp. n. (Tab. III. fig. 2.)

Dull obscure castaneous; head, anterior half of pronotum, and basal area of scutellum dull ochraceous; apex of scutellum levigate and pale olivaceous; connexivum luteous, spotted with blackish; membrane black, its apex paler; body beneath and legs very pale olivaceous. Body above thickly, darkly, and coarsely punctate; beneath much more sparsely punctate; femora with two small black spots near apex, and lateral margins of abdomen beneath with a series of small black segmental spots; rostrum just passing posterior coxæ with its apex black. Antennæ pale fuscous, basal joint (excluding apex) luteous; second and third joints subequal in length, or second slightly shorter than the third.

Long. 9 millim.

This species differs from *H. scapularis*, Thunb., by the spotted connexivum, and from *H. apicalis*, Herr.-Sch., it is distinguished by the more elongate body, different colour, &c. (*W. L. D.*)

Halyomorpha capitata, sp. n. (Tab. III. fig. 3.)

Body above ochraceous, thickly and irregularly covered with dark punctures. Head with the eyes fuscous, the ocelli red, and somewhat thickly covered with coarse brown punctures. Pronotum thickly and coarsely punctate, on each side of disk the punctures form obscure oblique fasciæ; lateral margins pale ochraceous and impunctate. Scutellum coarsely punctate, near lateral margins and before apex the punctures are confluent and castaneous, apex obscure pale olivaceous with scattered dark punctures; corium thickly, coarsely, and darkly punctate, castaneous in hue, excepting lateral margins, which are ochraceous; membrane purplish brown, with a submarginal tinge of black; connexivum ochraceous, with a double series of blackish spots at segmental margins. Body beneath and legs ochraceous, apical half of rostrum blackish, some small lateral sternal spots, smaller scattered spots on disk of abdomen, small stigmatal spots, and a series of marginal spots at segmental incisures black. Antennæ ochraceous, fourth and fifth joints and the apex of third fuscous, bases and apices of fourth and fifth joints ochraceous.

Long. 12–14 millim.

In this species the head is somewhat long and narrow, a character which will alone distinguish it from other species of the genus. (*W. L. D.*)

Halyomorpha pretoriæ, sp. n. (Tab. III. fig. 4.)

Above dull ochraceous, irregularly shaded with dark punctures; scutellum with the lateral and apical margins distinctly infuscated; corium with the disk more or less castaneous; membrane black, shining; connexivum with fuscous spots at segmental incisures; body beneath, rostrum, and legs ochraceous; rostrum with a central line and apex black; lateral margins of abdomen obscurely infuscated. Antennæ obscure brownish; second and third joints subequal in length and darkest in hue, fourth and fifth joints also subequal in length, fifth joint infuscated at centre.

Long. 12 millim. (*W. L. D.*)

Veterna patula, sp. n. (Tab. III. fig. 5.)

Body above ochraceous; basal area of pronotum from between lateral angles and corium (excluding lateral margins) purplish or olivaceous. Head with the eyes fuscous; pronotum with four black spots near anterior margin, the lateral angles black margined with carmine-red; scutellum with some clusters of dark punctures at base, and the same at lateral margins a little before apex, which is pale olivaceous; connexivum spotted with fuscous (sometimes immaculate); membrane black, shining. Body beneath and legs ochraceous, apex of rostrum, apices of pronotal angles, stigmata, and some lateral sternal spots black. Antennæ castaneous, basal joint luteous, fourth and fifth joints infuscated; second joint longest.

Long. 12 millim., exp. pronot. angl. 8. millim.

This species is allied to *V. pugionata*, Stål, by the shape of the pronotal angles, but is broader and without the white spots at base of scutellum. (*W. L. D.*)

Antestia transvaalia, sp. n. (Tab. III. fig. 6.)

Above dull ochraceous, somewhat thickly punctured with brown. Head with the margins of the central lobe enclosing two short central lines at base, the inner margins of the eyes and a cluster of punctures near the apex of each lateral lobe, black. Pronotum with the posterior area thickly covered with coarse black punctures and some scattered punctures on anterior area, the punctures form two obscure dark spots on disk; anterior and lateral margins and a central longitudinal discal line luteous and levigate, a black line on lateral margins near the posterior angles. Scutellum thickly covered with black punctures; two large spots at base and the apex luteous and almost impunctate; the black punctures become confluent near base, and form two obscure spots before apex. Corium thickly covered with black punctures, excepting at base of lateral margins and an angulated fascia on disk, both of which are luteous and levigate. Membrane black, its apical margin hyaline; connexivum luteous, spotted with black. Body beneath and legs luteous; abdomen with three basal creamy levigate fasciæ, the second and third interrupted at centre; rostrum with a central line and apex black; margins of abdomen spotted with black. Antennæ mutilated.

Long. 7 millim.

This species is allied to *A. variegata*, Thunb., from which the white levigate fasciæ on the under surface of the abdomen will alone render it very distinct. (*W. L. D.*)

Plinachtus falcatus, sp. n. (Tab. III. fig. 8.)

Body above and antennæ reddish ochraceous; head with two curved black lines extending from base to near emergence of antennæ; eyes dark fuscous; pronotum rugulose, lateral margins and apices of spines narrowly black, sub-anterior and sub-lateral margins and the spines (excluding apices) reddish, a central pale levigate line margined with black punctures, and with some scattered black punctures on basal area; scutellum darkly punctate, its apex black; corium very darkly punctate, its outer margins luteous and levigate, inwardly edged with black; membrane pale testaceous, exhibiting from beneath a black spot on each side. Body beneath, legs, and rostrum pale reddish ochraceous.

The second joint of the antennæ is longest, and the apical joint is somewhat infuscated.

Long. 11 millim.

This species may be distinguished by the pronotal spines being much less directed forwardly than is usual in the genus, and also by the two black spots seen through the membrane near apex of abdomen. (*W. L. D.*)

Lygæus planitiæ, sp. n. (Tab. III. fig. 7.)

Body above reddish orange, finely pilose; antennæ, eyes, central lobe to head, and a large basal spot at inner margin of eyes, two large wide discal longitudinal fasciæ to pronotum (strongly constricted anteriorly, fused near anterior margin, and connected with lateral margins at base and centre), scutellum excluding apex, apical half of clavus and claval margin, a large discal spot to corium connected with its lateral margin for half its length, membrane excluding lateral and apical margins, rostrum, coxæ, legs, margins and sutures of sternum, and sutural fasciæ to abdomen, black; margins of membrane pale fuscous.

Long. 10 to 12 millim. (*W. L. D.*)

Lygæus desertus, sp. n. (Tab. III. fig. 9.)

Closely allied to *L. planitiæ*, but differing by the following characters: corium with the lateral and apical margins, a broad claval marginal fascia, and a longitudinal fascia extending from the base where it is narrowest, to the centre of the apical margin where it is broadest, black. Membrane pale grey, hyaline, with an irregular reddish spot at the centre of basal margin, and a small obscure fuscous spot at inner apical margin.

Long. 10 millim. (*W. L. D.*)

Lygæus campestris, sp. n. (Tab. III. fig. 10.)

Body above reddish orange; antennæ, eyes, central lobe of head, inner margin of eyes, two irregular longitudinal central fasciæ to pronotum (which become somewhat evanescent posteriorly and are united to lateral margin a little beyond centre), lateral pronotal margins, scutellum excluding apical carina, a small spot near each apical margin of clavus, apical margin, posterior half of sublateral margin, and a spot on disk of corium, subbasal spot to membrane, sternum, legs, and rostrum, black. Membrane pale fuscous, the veins black, and a white spot on each side of the subbasal black spot. Extreme lateral margins of corium ochraceous; sternum with some large greyish-white spots.

Long. 8 millim. (*W. L. D.*)

Transvaalia, gen. nov.

Ocelli much nearer to eyes than to each other. Body elongate; head triangular and convex; pronotum excavated anteriorly, its lateral margins and a central longitudinal ridge carinate, basal margin with its apices angularly reflexed backwardly and inwardly. Rostrum with the basal joint extending a little beyond the base of the head. Scutellum triangular, moderately convex, and with a central longitudinal sulcation on basal half. Legs long; body above moderately pilose.

The peculiar structure of the pronotum and scutellum is sufficient to easily distinguish this genus. (*W. L. D.*)

Transvaalia lugens, sp. n. (Tab. III. fig. 12.)

Body above shining luteous; head, antennæ, legs, anterior area of pronotum, basal margin of scutellum, membrane, marginal sutures of pronotum, and apical half of abdomen (excluding segmental margins) black; a large furcate spot to head above and apex of head beneath, a spot at each anterior marginal area of pronotum, and coxal spot carmine-red.

Second joint of the antennæ longest, third and fourth subequal in length.

Long. 12 millim. (*W. L. D.*)

Nysius novitius, sp. n. (Tab. III. fig. 11.)

Body above ochraceous; head brownish, with the lateral margins, a central longitudinal fascia with a rounded spot on each side of disk ochraceous, the ochraceous markings margined with black; antennæ luteous, the basal and apical joints infuscated; eyes blackish. Pronotum coarsely covered with brown punctures, the lateral margins luteous and levigate, with a double curved linear mark on disk and a spot near posterior angle black. Scutellum black, with an irregular luteous marginal fascia on each side extending for about half the length from base. Corium with scattered coarse punctures, the lateral margins impunctate, and with three marginal brown spots, one below centre of lateral margin, and one at each apex of apical margin. Membrane pale ochraceous. Legs ochraceous, apices of femora fuscous, coxæ luteous; sternum strongly punctured with brown; abdomen beneath blackish.

Long. 5 millim. (*W. L. D.*)

Reduvius pulvisculatus, sp. n. (Tab. II. fig. 3.)

Body above purplish brown, thickly spotted with greyish pile, more thickly on corium than on pronotum. Head, antennæ, rostrum, anterior lobe of pronotum, scutellum, membrane, body beneath, femora, and tarsi black; tibiæ red, their bases and apices black; anterior lobe of pronotum thickly covered with ochraceous pile; head more sparingly pilose; connexivum above and beneath pale stramineous, slightly spotted with brownish at the segmental incisures.

Long. 20 millim.

Allied to *R. albopunctata*, Stål. (*W. L. D.*)

Reduvius sertus, sp. n. (Tab. II. fig. 8.)

Head, antennæ, rostrum, scutellum, membrane, body beneath, and legs black; a large quadrate spot on head extending from front of eyes to base of antennæ; pronotum, corium, connexivum above and beneath, coxæ, central area and lateral margins of sternum, and a discal patch to abdomen ochraceous.

Anterior lobe of pronotum irregularly and rugosely wrinkled; scutellum deeply and obliquely striated.

Var. *a*. Scutellum ochraceous.

Long. 17 millim. (*W. L. D.*)

Reduvius capitalis, sp. n. (Tab. II. fig. 1.)

Black; a large quadrate spot on head extending from front of eyes to base of antennæ, lateral areas of anterior lobe of pronotum, angular areas and basal margin of posterior lobe, scutellum (excluding basal angles), an angulated fascia to corium, connexivum above and beneath, head beneath, coxæ and coxal spots, parts of lateral margins of sternum, and abdomen beneath sanguineous; segmental spots to connexivum and margins of abdominal segments black; anterior femora annulated with sanguineous.

Long. 17 millim. (*W. L. D.*)

Pirates conspurcatus, sp. n. (Tab. II. fig. 10.)

Head, antennæ, rostrum, pronotum, scutellum, sternum, and legs black; corium and abdomen beneath ochraceous; base and apex of abdomen black; membrane, claval area, and lateral margins of corium fuscous; a black spot near inner angle of corium, a much larger black spot on disk of membrane, and a rectangular black spot on claval area. Abdomen above ochraceous.

Long. 10 millim.

Allied to *P. balteatus*, Germ. (*W. L. D.*)

NEUROPTERA.

The first signs of returning summer, with warmer nights and mornings, were shown by the appearance of Dragonflies hovering over the few small ponds to be found near Pretoria. The earliest to appear were *Orthetrum fasciculata* and *O. subfasciolata*, *Crocothemis erythræa*, and the gigantic *Anax mauricianus*, all these species being very abundant. *Tramea basilaris* is very rare and I only took or saw one specimen, whose wings were wet after heavy rains and floods. The genus *Palpopleura* I could only find represented by one species (*P. jucunda*) around Pretoria. Ant-lions also are somewhat scarce, frequenting the open veld, and I never saw but one species of *Palpares* (*P. caffer*) in the district of Pretoria, and, at all events, this is the dominant and representative species. The time and manner of appearance in the winged form of *Termes angustatus* has already been described (*ante*, p. 49).

I received much assistance from Mr. W. F. Kirby in identifying the species of Libellulinæ, and Mr. R. McLachlan kindly named two species of Ant-lions belonging to the collection.

NEUROPTERA-PLANIPENNIA.

Fam. MYRMELEONIDÆ.

Palpares caffer, Burm.	Pretoria.
Myrmeleon trivirgatus, Gerst.	Zoutpansberg.
Creagris flavipennis, Burm.	Pretoria.

PSEUDO-NEUROPTERA.

Fam. TERMITIDÆ.

Termes angustatus, Ramb.	Pretoria.
Termes, sp.?	Pretoria.

Fam. LIBELLULIDÆ.

Tramea basilaris, Ramb.	Pretoria.
Palpopleura portia, Dru.	Zoutpansberg.
Palpopleura lucia, Dru.	Durban, Natal.
Palpopleura jucunda, Ramb.	Pretoria and Zoutpansberg.
Sympetrum fonscolombii, Selys.	Pretoria and Zoutpansberg.
Trithemis dorsalis, Ramb.	Pretoria.
Trithemis sanguinolenta, Burm.	Pretoria and Zoutpansberg.
Crocothemis erythræa, Brullé.	Pretoria and Zoutpansberg.
Orthetrum fasciculata, Ramb.	Pretoria.
Orthetrum subfasciolata, Brauer.	Pretoria.

Subfam. ÆSCHNINÆ.
Anax mauricianus, Ramb. Pretoria.

ORTHOPTERA.

The vast plains of the Transvaal are peopled by myriads of insects belonging to this order, and when the summer is advanced it is impossible to walk across the veld without putting on wing numbers of the smaller species of the Acridiidæ, particularly the widely distributed *Acrida nasuta*, and species of the genera *Paracinema* and *Catantops*. It is this wealth of orthopterous life that provides the food for so many birds, and their disappearance at the dry season is also the principal reason of much bird migration in Southern Africa.

A few species are on the wing during the winter or dry season, and the largest and most brilliant example of these is *Acridium rubellum*, which also has the strongest flight of any orthopteron I saw, and when on the wing may easily be mistaken for a small bird. The next prominent insect to appear is *Phymateus leprosus*, which has a very sluggish flight, is easily captured, and precedes its generic companion *Phymateus morbillosus*. The most active and high-flying species of *Phymateus* is *P. squarrosus*, which increases in numbers as the summer progresses, when the varieties with the pronotum concolorous and that with the margins red are both found together. Another brilliant species which I found on the barest and most sterile veld is *Œdalus acutangulus*, whose blue and black wings make it very conspicuous during flight. The vast swarms of *Pachytylus migratoroides* which visited the Transvaal during my stay have already been recorded (*ante*, p. 71), and in the Locustidæ the peculiar habits of *Clonia wahlbergi* and *Hemisaga prædatoria* have also been described (*ante*, pp. 83 & 65).

I am much indebted to Mons. Henri de Saussure for much valuable assistance in the identification of the species of this order.

Fam. FORFICULIDÆ.

Labidura riparia, Pall. Pretoria.
Labidura, sp.? Pretoria.

Fam. BLATTIDÆ.

Ectobia ericetorum, Wesm.	Pretoria.
Phyllodromia germanica, Linn.	Pretoria.
Deropeltis erythrocephala, Fabr.	Pretoria.
Nauphœta circumvagens, Burm.	Pretoria.
Derocalymma capucina, Gerst.	Pretoria.

Fam. PHASMIDÆ.

Clonaria? guenzii, Bates.	Durban, Natal.

Fam. MANTIDÆ.

Pyrgomantis singularis, Gerst.	Pretoria.
Lygdamia capitata, Sauss.	Pretoria.
Parathespis galeata, Gerst.	Pretoria and Zoutpans-
Miomantis fenestrata, Fabr.	Pretoria. [berg.
Mantis sacra, Thunb.	Pretoria.
Hierodula gastrica, Stål.	Pretoria.
Harpax tricolor, Linn.	Pretoria.
Phyllocrania paradoxa, Burm.	Pretoria.

Fam. GRYLLIDÆ.

Gryllotalpa africana, Pal. Beauv.	Pretoria.
Gryllus bimaculatus, De Geer.	Pretoria.
Gryllus leucostomus, Serv.	Pretoria.
Gryllus compactus, Walk.	Pretoria.
Œcanthos capensis, Sauss.	Pretoria.

Fam. LOCUSTIDÆ.

Conocephalus mandibularis, Charp.	Pretoria.
Clonia wahlbergi, Stål.	Waterberg.

Hemisaja prædatoria, Distant, sp. n. (fig. *ante*, p. 63).

♀. Ochraceous, disk of abdomen and ovipositer testaceous, apical area of abdomen greenish; head with two faint black central longitudinal lines, continued more distinctly along the whole length of the pronotum and abdomen. Lateral furrowed margin of pronotum white and somewhat bifasciate, with its inner edge narrowly black; lateral margins of meso- and metanotum white; lateral margins of abdomen with a narrow white fascia, inwardly testaceous and outwardly black. Wings quite rudimentary and greenish ochraceous.

Anterior femora a little longer in length than head and pronotum together and armed with ten spines on each side; anterior tibiæ with eight spines on each side; intermediate femora with nine spines on inner under margin, and ten shorter spines on outer under margin; intermediate tibiæ with ten or eleven spines on each side; posterior femora with a double series of short spines beneath, but not reaching apical area.

Long. from apex of head to apex of ovipositor 76 millim.; pronot. 7; head 5; ovipositor 28; ant. femora 13; post. femora 38.   Pretoria.

Differs from *H. hastata*, the only other described species of the genus, both as regards the different markings and the proportional measurements.

Fam. ACRIDIIDÆ.

Acrida nasuta, Linn.	Pretoria.
Acrida turrita, Linn., var. *calæata*, Sauss.	Pretoria.
Mesops abbreviatus, Pal. Beauv.	Zoutpansberg.
Parga gracilis, Burm.	Pretoria.
Phymateus leprosus, Fabr.	Pretoria.
Phymateus morbillosus, Linn.	Pretoria.
Phymateus squarrosus, Linn.	Pretoria.
Ochrophlebia ligneola, Serv.	Pretoria.
Maura rubro-ornata, Stål.	Pretoria.
Taphronota porosa, Stål.	Pretoria.
Petasia spumans, Thunb.	Pretoria.

—— ——, var. *ater*, Dist. (Tab. IV. fig. 3.)

Black; anterior and posterior margins of thorax above, interior area of wings, basal halves of anterior and intermediate femora, tibiæ excluding bases and apices, tarsi excluding apices, and the palpi bright coralline red.   Pretoria.

Porthetis cinerascens, Stål.	Pretoria.
Porthetis griseus, Serv.	Pretoria.

Xiphocera distanti, sp. n.	Pretoria.
Xiphocera picta, sp. n.	Waterberg.
Acridium ruficorne, Fabr.	Pretoria.
Acridium hottentottum, Stål.	Pretoria.
Acridium tataricum, Linn.	Zoutpansberg.
Acridium rubellum, Serv.	Pretoria and Zoutpansberg.
Catantops humeralis, Thunb.	Pretoria.
Catantops sulphureus, Walk.	Durban, *Natal*.
Calliptenus crassus, Walk.	Pretoria.
Calliptenus pallidicornis, Stål.	Pretoria.
Pezotettix capensis, Thunb.	Pretoria.
Chrotogonus meridionalis, sp. n.	Zoutpansberg.
Paracinema tricolor, Thunb.	Pretoria.
Paracinema basalis, Walk.	Pretoria.
Pycnodictya obscura, Linn.	Pretoria.
Pycnodictya adustum, Walk.	Pretoria.
Cosmorhyssa fasciata, Thunb.	Pretoria.
Pachytylus migratoroides, Reiche.	Pretoria.
Pachytylus lucasii, Brunn.	Pretoria.
Œdalus plena, Walk.	Pretoria.
Œdalus marmoratus, Thunb., var.	Pretoria.
Œdalus acutangulus, Stål.	Pretoria.
Œdalus nigro-fasciatus, Charp.	Pretoria.
Œdalus citrinus, Sauss.	Zoutpansberg.
Humbella tenuicornis, Schaum.	Pretoria.
Heteropternis hyalina, Sauss.	Zoutpansberg.
Tmetonota sabra, Sauss.	Pretoria.
Trilophidia annulata, Thunb.	Pretoria.
Acrotylus flavescens, Stål.	Pretoria.

Descriptions.

(By Mons. H. DE SAUSSURE.)

Familia PAMPHAGIDÆ.

Genus *Xiphocera*, Latr., Stål, Saussure.

The two species which follow belong to the group of *X. mannulus*, Saussure, having filiform antennæ, not dilated.

(Comp. H. de Saussure, 'Spicilegia entomologica Genevensis,' ii. 1887, p. 33.)

They differ from that species and from others of the same group by their strongly compressed body, and can be distinguished from one another as follows:—

a. The crest of pronotum having three ovoid pellucid fenestræ, very distinct; the crest entire to its apex.—*X. distanti*.
b. The crest of pronotum having but small pellucid fenestræ, obsolete; the crest crenulated at its apex.—*X. picta*.

Xiphocera distanti, Saussure, sp. n. (Tab. IV. fig. 1.)

Fusco-grisea, valde compressa, crasse granulata. Antennæ filiformes, 15-16-articulatæ, superne totæ obsolete, subtus articulis 6 primis planatæ. Articuli omnes depressi, extus acuti, facie interna angusta subrotundata, haud marginata; articuli 7 apicales submoniliformes.

Caput confertim, inter carinas laterales sparse, granulatum. Vertex vix declivis, a latere rectangulatus; ejus scutellum breve, excavatum, antice rectangulatum, marginibus valde prominulis, postice rotundatum. Ejus carinulæ laterales incisæ, angulis incisuræ rotundatis, haud crenatis. Costa facialis a latere parum incisa.

Pronotum utrinque inferius dense granulosum. Crista elevata, lamellaris, supra occiput valde acutangulatim producta (Sauss. *l. c.* pl. ii. fig. 7); ejus linea dorsalis a latere in medio vix, antice et postice magis arcuata (Sauss. *l. c.* pl. iii. fig. 23); ejus pars postica valde producta, nec sinuata nec crenata, apice acuto, bidentato. Sulci 3 in crista profundi ac fenestrati, impressiones profundas, ovatas, diaphanas majores obferentes. Pars dorsalis pronoti verruculosa ac sparse acute granosa.—Pedes graciles, antice haud granulati, sed extus carinati. Femora postica haud late dilatata, ante concha apicali coarctata (Sauss. *l. c.* pl. iii. fig. 20), cristis parum latis, supra 9-spinosa, infra irregulariter denticulata.

Abdominis segmenta 1^m–6^m superne ante apicem spina trigonali armata.

♀. Long. 46 millim., pronot. 16, fem. post. 16.

Pretoria, 1 ♀.

Xiphocera picta, Saussure, sp. n. (Tab. IV. fig. 2.)

Statura et forma sensim ut in *X. distanti*; corpore toto sabuloso, excepto in parte apicali abdominis; colore fusco, valde albo et flavido strigato.—Antennæ 16-articulatæ, deplanato-prismaticæ, supra planæ, subtus articulis 3°–8° in medio carinatis ; 7 ultimis submoniliformibus, angulo externo prominulo.—Scutelli verticis margines laterales incisi ac crenulati.—Pronoti crista quam in *X. distanti* minus arcuata, apice crenata. Sulci 3 in illa minus impressi, lineares, fenestris parum translucidis, minoribus. Pars dorsalis tuberculis minutis dentiformibus conspersa; crista confertim granulata vel verruculosa, rugosissima; lobi laterales sparse nigro-verruculati, margine postico minute denticulato.—Femora postica albida, fusco 3-fasciata, necnon area inferiore albo 4-maculata; margo superior spinis trigonalibus fortibus armatus, genu spina apicali armato; margo inferior valde denticulatus.—Segmenta abdominis omnia superne dente acuto præapicali armata.

Pictura : Facies albo et nigro marmorata. Corpus utrinque linea albida callosa, interrupta, percurrente, ornatum. Lobi laterales pronoti lineis albis irregularibus, crista et dorsum vittis albis obliquis, descendentibus, notata. Metathorax utrinque inferius vittis 2 albis et superius vitta obliqua in 1° abdominis segmento perducta. Corpus subtus pallidum.

♀. Long. 36 millim., pronot. 16, fem. post. 15.

Waterberg.

Familia PYRGOMORPHIDÆ.

Chrotogonus meridionalis, Saussure, sp. n. (Tab. IV. fig. 5.)

Gracilis, fulvo-griseus, fusco-umbratus; pedibus fusco-fasciatis. Costa facialis a latere visa tenuiter sinuata, haud angulatim incisa. Verticis rostrum sat prominulum, parabolicum, oculos plus quam dimidia latitudine superans.—Pronotum margine postico transverso, arcuato, haud lobato, tuberculis nigris 5. Prozona supra tenuiter tuberculosa; ejus partes 3 æquales. Metazona tantum rare tuberculata ac subtiliter granulata. Lobi laterales fere rectangulatim deflexi, angulo postico rectangulo vel leviter retro-producto, haud lobato.—Elytra angustissima, abdominis apicem attingentia, nigro-punctata, seriebus tuberculorum nullis; area costali parum dilatata. Alæ nigræ.

♀. Long. 21 millim., pronot. 4, fem. 9, elytr. 15, latit. 3.

This species is characterized by its narrow form, by its elytra being narrow and not tuberculated, and by its pronotum having the hind border not angulated, but only arcuated, and less than in *Chr. hemipterus*.

Zoutpansberg.

INDEX.

Abacetus obtusus, 188.
Acacia mollissima, 115.
Acanthogenius dorsalis, 188.
Acanthocoris lugens, 246.
Acanthorrhinus dregei, 200.
Acaridæ, 126, 179.
    Found under stones in Transvaal, 179.
    Loss to live-stock in Natal, 126.
Accidents common to all animal life, 55.
Achænops facialis, sp. n., 204.
Acharana otreusalis, 241.
Acherontia atropos, 18, 68, 153, 236.
Acidalia gazella, 240.
Acræa anemosa, 233.
—— *buxtoni*, 233.
—— *caldarena*, 233.
—— *doubledayi*, 233.
—— *encedon*, 233.
—— *horta*, 232.
—— *natalica*, 123, 233.
—— *neobule*, 233.
—— *nohara*, 233.
—— *rahira*, 233.
—— *violarum*, 233.
Acrida nasuta, 259.
—— *turrita*, var. *calcata*, 259.
Acridium hottentottum, 260.
—— *rubellum*, 260.
—— *ruficorne*, 260.
—— *tataricum*, 260.
Acrotylus flavescens, 260.
Adalia flavomaculata, 209.
Adoretus luteipes, 47, 192.
Ægialitis asiaticus, 169.
—— *tricollaris*, 169.

Ænidea pretoriæ, sp. n., 206, 209.
Æthriostoma gloriosa, 190.
Agama atricollis, 174.
—— *hispida*, 87, 173, 174.
    Habits under stones, 87.
Aglossa basalis, 241.
Agonoscelis erosa, 246.
—— *puberula*, 246.
—— *versicolor*, 246.
Agrotis biconica, 238.
Alæmon nivosa, 168.
Alæna amazoula, 234.
Alcides senex, 200.
Alesia inclusa, 209.
Aliteus adspersus, 195.
Aloes, 45.
Amblyomma hebræum, 179.
Amblysterna vittipennis, 195.
Amiantus gibbosus, 190.
—— *undosus*, sp. n., 199.
Ammophila bonæ spei, 211.
Amphidesmus analis, 196, 201.
Ampulex nigro-cærulea, sp. n., 211, 212.
Amydrus morio, 164, 167.
Anace lateritia, 237.
Anax mauricianus, 45, 257.
Ancylopus fuscipennis, sp. n., 210.
Anisorhina flavomaculata, var. *egregia*, 194.
Anomalipes complanatus, 190.
—— *intermedius*, 199.
—— *talpa*, 199.
—— *variolosus*, 199.
Anoplochilus tomentosus, 194.
Anoplocnemis curvipes, 46, 246.
Anteris zelleri, 236.

Antestia transvaalia, sp. n., 246, 251.
Anthia æquilatera, 188.
—— *desertorum*, 188.
—— *maxillosa*, 50, 187, 188.
—— *mellyi*, 188.
—— *thoracica*, 50, 187, 188.
Anthocomus, sp., 198.
Anthropoides paradisea, 50, 108.
Antipus rufus, 204.
Ant-lion, 91.
Anubis mellyi, 201.
Aphelia apolinaris, 123, 238.
Aphodius lividus, 152, 191.
Aplasta dichroa, 194.
Apples in Natal, 126.
    Devastated by Cetoniid beetles, 115, 126, 193.
Aquila wahlbergi, 163, 165.
Arachnida, 179.
Araneæ, 180.
Argentiferous wealth near Pretoria, 59.
Argiope nigrovittata, 180.
Arthileta cloeckneria, 236.
Asbecesta cyanipennis, 206.
Asio capensis, 112, 165.
    Feeding on large coleoptera, 112.
Aspilocoryphus fasciativentris, 247.
Aspongopus japetus, 246.
—— *nubilus*, 246.
Astur polyzonoides, 165.
Atella columbina, 233.
Athalia bicolor, sp. n., 211, 226.
Atractonotus formicarius, 187, 188.
Attagenus, sp., 190.
Atylotus, sp., 244.
Aulacophora vinula, 206.
Aulonogyrus abdominalis, 45, 189.
Auriferous deposits absent at Pretoria, 50.
Austen, Ernest E., 244.
"Australian bug," 88.
Aves, 152, 163.
    The Transvaal avifauna already well worked, 163.
Axylia cinctothorax, 233.

Azanus sesous, 234.

Bagrada hilaris, 246.
Bamboos, 121.
Bananas, 121, 125.
Basipta stolida, 209.
Bates, H. W., ix, 41, 187.
Beach Wood at Durban, 123.
Belonogaster rufipennis, 210.
Berea at Durban, 122.
"Berg bas," 43.
Biggarsberg, 9.
Binna linea, 237.
—— *madagascariensis*, 70, 237.
    Attacked by Cape Wagtail, 70.
Birds of prey around Pretoria, 55.
    Their extremely fat condition, 57.
Blue Cranes, 50.
Bluff at Durban, 122.
Boers, 20.
    As soldiers, 32.
    Farmers, 21, 22.
    Physical appearance, 35.
    Sobriety, 36.
    Theology, 23, 26, 33.
    Treatment of Kafirs, 24.
    Want of gaiety, 36.
Boers and English, 35, 90, 147, 148.
Boer War, 10, 31, 32, 140.
Bolboceras batesii, sp. n., 191.
Boom, great, the, 145.
Boulenger, G. A., 173.
Bourgeois, Mons. J., 196.
Bowker, Colonel, 41, 121.
Brachinus armiger, 188.
Brachycerus apterus, 200.
—— *cancellatus*, 200.
—— *natalensis*, 200.
Bracon fastidiator, 211.
—— *flagrator*, 211.
Bradyornis silens, 167.
Bubulcus ibis, 169.
Bufo carens, 176.
—— *regularis*, 48, 176.

Buteo desertorum, 56, 163, 165.
—— *jakal*, 112, 163, 165.
Butler, A. G., 235.
Butterflies, 5, 39, 40, 46, 61, 91, 112, 121, 123.
Buzzards, 57, 80, 112.
  Instinctive terror of these birds possessed by poultry, 57.

Cænina pæcilaria, 240.
Cænomorpha nervosa, 245.
Calleida angustata, 188.
Callidea natalensis, 245.
Calliphora marginalis, 244.
Calliptenus crassus, 260.
—— *pallidicornis*, 260.
Callistomimus sexpustulatus, 187, 188.
Calomorpha wahlbergi, 205.
Calpe apicalis, 237.
Camponotus grandidieri, 211.
Camptolenes cribraria, 204.
Candèze, Dr. E., 195.
Canis mesomelas, 111.
Canna indica, 18.
Cantharsius, sp., 191.
Cape Town, 4.
  Malay population, 5.
  Museum, 5.
Carbula trisignata, 246.
Cardiophorus præmorsus, 195.
Carebara vidua, 211.
Cassida hybrida, 209.
—— *lurida*, 209.
—— *punctata*, 209.
—— *scripta*, 209.
Castalius melæna, 234.
Castellated residence in Zoutpansberg, 96.
Catantops humeralis, 260.
—— *sulphureus*, 260.
Catochrysops osiris, 233.
Catopsilia florella, 46, 234.
Caura ruficentris, 246.
Causus rhombeatus, 176.
Cephalolophus grimmii, 111, 159.

Cerchneis amurensis, 112, 165.
  Feeds on Orthoptera, 112.
—— *rupicola*, 128, 165.
—— *rupicoloides*, 165.
—— *tinnunculoides*, 112, 165.
  Feeds on Orthoptera, 112, 164.
Ceroplesis bicincta, 202.
—— *brachyptera*, 202.
—— *capensis*, var. n., 202, 203.
Cervicapra arundinum, 159.
Cetonia cincta, 194.
Cetoniids, 66, 92, 123, 193.
Chalcogenia cuprea, 195.
Chamæleon parvilobus, 174.
Champion, J. C., 198.
Charlestown, 119, 120.
Chera progne, 50, 167.
  Nuptial plumage a hindrance to flight, 50.
  Sexual selection exercised at the expense of protection, 164.
Chettusia coronata, 169.
Chlænius cylindricollis, 188.
—— *subsulcatus*, 188.
—— *vitticollis*, 187, 188.
Chlorion xanthocerum, 210.
Chærocampa nerii, 18.
—— *schenki*, 236.
Chrotogonus meridionalis, sp. n., 260, 262.
Chrysis (*Pyria*) *lynica*, 211.
Chrysococcyx cupreus, 164.
Chrysomela opulenta, 206.
Chrysorychia amanga, 234.
Cicadas, 67.
Cidaria pudicata, 240.
Cigaritis leroma, 234.
Cimex figuratus, var. n., 245, 248.
Circus pygargus, 152, 163, 165.
Cleonus, sp., 200.
Cletus ochraceus, 246.
Clivina grandis, 188.
Clonaria? guenzii, 258.
Clonia wahlbergi, 84, 258.
Clytanthus capensis, 201.
Clythra wahlbergi, 204.

Coaching, 9.
Coccinellidæ, 66, 209.
Colasposoma pubescens, 205.
Coleoptera, 152, 187.
    After the rains, 50.
    Difficult to obtain, 92.
    On leaves after rain, 123.
    Under stones, 39.
Colias electra, 30, 234.
Colley, General, his grave, 11.
Colonial settlers in the Transvaal, 134.
    Their great value to the State, 135.
Colonist, an old, 90.
Colotes, sp., 198.
Colours of animals fixed by natural selection during the great struggle for existence subsequent to the last geological change of surface, 76.
Colpoon compressum, 42, 77.
Columba phæonota, 168.
Commercial morality, 137.
—— prosperity of the Transvaal retarded by want of railway communication, 130, 138.
Company promoter, the, 127, 138.
Comparison of social life and politics in the Transvaal and Britain, 86.
Compsomera elegantissima, 201.
Conocephalus mandibularis, 258.
Convolvulus, a fine, 96.
Copridæ, their flight, 190.
Copris contractus, 191.
—— *fidius*, 191.
Coptomia umbrosa, 123, 193.
Coptorhina klugii, 191.
Coracias caudata, 80, 112, 166.
Corvus scapulatus, 96, 167.
Corynetes ruficollis, 152, 198.
—— *rufipes*, 152, 198.
Corynodes compressicornis, 205.
Cosmorhyssa fasciata, 260.
Cossypha caffra, 106.

Crania, belonging to the Makapan tribe, descriptions of, 157.
Creagris flavipennis, 256.
Crepidogaster bimaculatus, 187, 188.
Crioceris constricticollis, 204.
—— *puncticollis*, 204.
Crocidura martensii, 159.
—— *pilosa*, 159.
Crocodile, mythical reports as to, 78.
Crocothemis erythræa, 45, 256.
Crossotus klugii, sp. n., 202, 203.
Crothæma decorata, sp. n., 237.
Cryptocephalus decemnotatus, 204.
—— *dregei*, 204.
—— *pardalis*, 204.
—— *pustulatus*, 204.
Cursorius chalcopterus, 169.
—— *senegalensis*, 169.
Cyclopides willemi, 235.
Cydonia lunata, 209.
—— *quadrilineata*, 209.
Cyligramma latona, 238.
Cynictis penicillata, 159.
Cypholoba ranzanii, 187, 188.
Cyphonistes vallatus, 193.
Cyphononyx antennata, sp. n., 211, 217.

Danais chrysippus, 39, 61, 232.
    Devoured by *Hemisaga prædatoria*, 65.
    Its varieties *alcippus* and *dorippus* found at Pretoria, 65, 232.
Decimia bicolorata, 237.
Deiopeia pulchella, 68, 153, 237.
    In flight mistaken for a *Lycænid*, 68.
    Its wide distribution, 68.
Dendritic or arborescent markings in quartzite, 58.
Dendropicus cardinalis, 106.
Dermantinus limbifer, 247.
—— *lugens*, 247.
Dermestes vulpinus, 152, 190.
Derocalymma capucina, 258.

Deropeltis erythrocephala, 258.
Destruction of birds by floods, 91.
Diceros algoensis, 193, 194.
—— ——, var. *flavipennis*, 193, 194.
Dichelocephala lanceolata, 246.
Dichelus vulpinus, 192.
Dichtha cubica, 198, 199.
Dicuches armipes, 247.
—— *patruelis*, 247.
Diplognatha hebræa, 50, 194.
—— *silicea*, 194.
Diptera, 46, 67, 244.
Discolia caffra, 211.
—— *præcana*, sp. n., 211, 222.
—— *præstabilis*, sp. n., 211, 222.
Dismegistus fimbriatus, 245.
Distantella, gen. n., 229.
—— *trinotata*, sp. n., 211, 230.
Dopper Church, 27.
Dorylus helvolus, 211.
Dragonflies, 45, 54, 112.
Dresser. H. E., 165.
Dromica gigantea, 187.
Durban, 8, 121.
    Beach Wood, 123.
    Berea, the, 122.
    Bluff, the, 122.
    Museum, 121.
    Pleasures of the naturalist at, 124.
Durrant, J. Hartley, 235, 242.
Dust-storm, 47.
"Duyker," the, 111.
Dwaas River, 97, 112.
Dynamite, 58, 89, 120.

East London, 8.
Ectobia ericetorum, 153, 258.
Edocla quadrisignata, 248.
Egybolia vaillantina, 122, 237.
Elanus cæruleus, 163, 165.
Elaphinis irrorata, 123, 194.
—— *latecostata*, 123, 194.
Elasmopoda undata, 246.
Eletica rufa, 198, 199.
Elis barbata, 211, 223.

Epilachna bifasciata, 209.
—— *dregei*, 209.
—— *hirta*, 209.
Epizeuxis æthiops, sp. n., 239.
Equus quagga, 75.
Eremias lineo-ocellata, 174.
Eriesthis guttata, 192.
—— *semihirta*, 192.
Eronia leda, 123, 234.
Estrelda astrild, 167.
Eubolia dvercana, 240.
—— *proxaulkaria*, 240.
—— *punicaria*, 240.
Eucalyptus, 17.
Euchromia africana, 123, 236.
Euclasta warreni, sp. n., 241.
Euleptus caffer, 188.
Eumenes tinctor, 210.
Euphorbia, 45, 80.
    Poisonous qualities, 80.
Eupoda, 203.
    Habits, 203.
Euporus callichromoides, 201.
Euryope terminalis, 205.
Eurytela hiarbas, 233.
Evochomus nigromaculatus, 152, 209.
Erocatus lineatus, 3.

Falco ruficollis, 164, 165.
Farm, Natal model, 126.
Farmer, British, wanted in the Transvaal, 138.
Farmers, Boer, 21, 22, 34, 45, 113.
    Great want of education, 133, 138.
Floods, and loss of life, 53, 60, 78, 88, 120.
Flowers, common English, cultivated in the Transvaal, 18.
Flying-fish, 3.
Francolinus gariepensis, 112, 168.
—— *levaillantii*, 75, 112, 168.
    Protective coloration, 75.
—— *subtorquatus*, 112, 168.

Gahan, C. J., 201, 204, 206.

Galerucidæ, 66, 206.
Gametis balteata, 194.
Gastracantha, sp., 180.
Gellia anguticollis, 246.
Geological features around Pretoria, 58.
Gerrhosaurus flavigularis, 174.
Gill, G. D., 94.
Gladstone, Mr., name of, in the Transvaal, 140.
Glareola melanoptera, 169.
Glaucidium perlatum, 165.
Glauconia distanti, sp. n., 175.
Glypsus conspicuus, 245.
—— *mœstus*, 245.
"Gom Paauw," 74.
Gomphocarpus, sp., 62, 66.
Gonodela amandata, 240.
Gorgophis libania, 238.
Gorham, Rev. H. S., 197.
Graphipterus cordiger, 188.
—— *ovatus*, 188.
—— *westwoodi*, 188.
Grass-fires, 21, 46, 114.
Gryllotalpa africana, 258.
Gryllus bimaculatus, 258.
—— *compactus*, 258.
—— *leucostomus*, 258.
Guinea-fowl, 13, 72.
Gymnopleurus cælatus, 191.
—— *wahlbergi*, 191.
Gynandrophthalma anisogramma, 204.
Gynanisa maia, 238.
Gyps kolbii, 69, 163, 165.

Hailstones forming blocks of ice, 53.
Halitonoma epistomata, 204.
Halyomorpha capitata, sp. n., 245, 249.
—— *pretoriæ*, sp. n., 246, 250.
Hamanumida dædalus, 40, 41, 233.
Haritala quaternalis, 241.
Harpalus capicola, 188.
Harpax tricolor, 258.
Hecyrida terrea, 202.

Hedybius amœnus, sp. n., 197.
Heidelberg, 117.
Heliocopris hamadryas, 191.
Helioryctes melanopyrus, 210.
Hemaris hylas, 236.
Hemipimpla, 227.
—— *caffra*, sp. n., 211, 227.
—— *calliptera*, sp. n., 211, 228.
Hemisaga prædatoria, sp. n., 65, 258.
    Devours *Danais chrysippus*, 65.
Herpænia eriphia, 91, 234.
Hesperia forestan, 235.
Heterocera, 235.
    Habits &c., 230.
Heterocorax capensis, 167.
Heteroderes inscriptus, 195.
Heteromera, 67, 198.
    Habits of, 198.
Heteronychus, sp., 193.
Heteroptera, 244.
Heteropternis hyalina, 260.
Hibiscus, 18, 125.
Hierodula gastrica, 258.
Himatismus buprestoides, 199.
Hindu race at Durban, 8.
Hippobosca rufipes, 244.
Hipporhinus corniculatus, 200.
—— *cornutus*, 200.
—— *pilularius*, 200.
Hirundo cucullata, 167.
—— *semirufa*, 167.
Hister caffer, 190.
—— *fossor*, 190.
—— *hottentotta*, 190.
—— *ovatula*, 190.
Holcostethus obscuratus, sp. n., 245, 249.
Hollanders, 35, 134, 135, 148.
Homœocerus annulatus, 246.
—— *magnicornis*, 246.
Homonotus cærulans, sp. n., 211, 213.
—— *pedestris*, sp. n., 211, 214.
Homoptera, 243.
Honey poisonous, when derived from bloom of Euphorbias, 80.
Hoplostomus fuligineus, 194.

Hoploterna valga, 246.
Horse-sickness, 95.
Humbella tenuicornis, 260.
Hybosorus arator, 191.
Hydrous, sp., 189.
Hymenoptera, 67, 210.
Hypanis ilithyia, 46, 233.
Hyperacantha oculata, 206.
Hyphantornis relatus, 167.
Hypolimnas misippus, 65, 233.
Hypolithus, sp., 188.
Hypselogenia concava, 193.
Hystrichopus caffer, 188.

Icerya purchasi, 88.
Illicit diamond-buying, 138.
Insects protected from their enemies by mimicry or protective resemblance frequently thus survive some incipient mortality which would have otherwise helped to effect their extinction, 62.
Iolaus bowkeri, 234.
Iron-work among the Mavenda Kafirs, 107.
Ischnostoma nasuta, 193.
Islam, a Priest of, 117.

Jackal, the, 111.
Jacoby, Martin, 204.
Janson, Oliver, 193.
Jews in the Transvaal, 86, 135.
    Financial ability, 136.
    Their cosmopolitanism and gaiety, 137.
    Their knowledge of the country, 136.
Johannesburg, 13, 116.
    Absence of trees, 116.
    Hotels, 117.
    Most English town in the Transvaal, 138.
"Jumping hare," 75.
Junonia boöpis, 233.
—— *cebrene*, 40, 233.
—— *clelia*, 233.

Kafir and Boer, 22, 24, 25, 81, 98.
Kafir labour, 22, 24, 98, 104, 140.
    Distance travelled to work, 99.
    Manual labour at present dependent on, 141.
    Work to procure money to obtain another wife, 99.
Kafir stores and Kafir traders, 97.
    Solitary life, 97.
Kestrels, 112, 163.
Kirby, W. F., 235, 256.
Krüger, S. J. P., President, 28, 33, 82.

Labidura riparia, 153, 257.
Ladysmith, 121.
Lælia adspersa, 237.
Lagria, sp., 199.
Laings Nek, 11, 120.
    Railway tunnel, 120.
Lamprocolius sycobius, 164, 167.
Lamprophis rufulus, 176.
Laniarius atrococcineus, 167.
—— *gutturalis*, 167.
Lanius collaris, 55, 167.
—— *collurio*, 167.
Larra ornata, 210.
Lebeda aculeata, 237.
Lebia, sp., 188.
Lepidoptera, 153, 231.
Leptodira rufescens, 176.
Leptosoma apicalis, 123, 236.
Leucania amens, 238.
—— *apparata*, 238.
—— *percussa*, 238.
—— *substituta*, 238.
Lewis, George, 190.
Life on the table-lands compared with that on the sea, 59.
Lion, probably in the Transvaal confined to Zoutpansberg, 157.
Liquor, love of, by Kafirs, 102.
Lithosia ? fumeola, 237.
Litopus dispar, 201.
Lixus, sp., 200.
Locris arithmetica, 248.
—— *transversa*, 248.

Locust chased by dogs, 39.
—— eaten by Kafirs, 72.
—— eaten by poultry and bustards, 72.
—— impaled on barbed wire, 55.
—— swarm, 71.
Longicornia, 201.
  Habits &c., 201.
Lophoceros leucomelas, 164, 166.
Lophonotus, sp., 244.
Lost in the wood-bush, 91.
Luciola capensis, 197.
Lycænesthes liodes, 234.
Lycauges donovani, sp. n., 240.
Lycus æolus, 196.
—— *ampliatus*, 196.
—— *bremei*, 123, 196.
—— *constrictus*, 196.
—— *dilatatus*, 196.
—— *distanti*, sp. n., 196.
—— *integripennis*, 196.
—— *kolbei*, 196.
—— *pyriformis*, 196.
—— *rostratus*, 196.
—— *subtrabeatus*, 196.
—— *zonatus*, 196.
Lygæidæ, 45, 67, 247.
Lygæosoma villosula, 247.
Lygæus campestris, sp. n., 247, 253.
—— *desertus*, sp. n., 247, 252.
—— *elegans*, 247.
—— *planitiæ*, sp. n., 247, 252.
—— *rivularis*, 247.
—— *septus*, 247.
—— *trilineatus*, 247.
Lygdamia capitata, 258.

Mabuia striata, 174.
—— *trivittata*, 87, 173, 174.
  Great resemblance to snake when curled up, 87.
  Inhabit holes with toads, 87.
Machetes pugnax, 169.
Machla porcella, 199.
MacLachlan, R., 256.
Macroglossa trochilus, 236.

Macroma cognata, 194.
Macronyx capensis, 164, 168.
Magato, chief of the Makatese, 107.
Magwambas, the, 100.
  Dress, 101.
  Funeral dances &c., 102.
  Killing an ox and great dance, 102.
  Love of strong liquor, 102.
  Number in the Spelonken, 101.
  Originally refugees from surrounding districts, 100.
Maiden-hair ferns, 45.
Maize cultivated by Kafirs, 83.
Majuba Hill, 10, 90, 120, 128, 135.
Makapan's Cave, 82, 84.
  Immense slaughter of Kafirs, 82.
  Kafirs blockaded by Boers, 82.
Makapan's Poort, 81.
  Scene of murder by Kafirs in 1854, 81.
Makatese, the, 107.
  Location in the Spelonken, 107.
Malacodermata, 196.
Malays at Cape Town, 5.
Mamestra breviuscula, 238.
—— *renisigna*, 238.
Mammalia, 152, 157.
Mangoes, 125.
Man's development of the Transvaal a struggle with the different forces and agents of Nature, 89.
Manticora tuberculata, 50, 187.
Mantis sacra, 258.
Maritzburg=Pietermaritzburg, 121, 127.
Mashonaland trek, 85.
Maura rubro-ornata, 259.
Mavendas, the, 107.
  Iron-smelting and iron manufacture, 107.
  Making a pick or hoe, 108.
Megachile maxillosa, 210.
Megalonychus interstitialis, 188.
Melanitis leda, 232.

Melinesthes umbonata, 194.
Melittophagus meridionalis, 103.
Melybœus crassus, 195.
Men who early in life visit South Africa seldom finally leave it, 87, 138, 139.
Menius distanti, sp. n., 205.
Merops apiaster, 44, 152, 166.
Mesa diapherogamia, sp. n., 211, 225.
Mesops abbreviatus, 250.
Micranterus validus, 199.
Microstylum dispar, 244.
Milvus ægyptius, 56, 165.
Mimicry, 41, 66, 92.
Mimosa-bark, 116.
Miomantis fenestrata, 258.
Mirperus jaculus, 246.
Mist and fog, 116.
Monitor, the, 77, 87, 174.
Monochelus, sp., 192.
Monolepta flaveola, 206.
Monticola brevipes, 166.
Morægamus globiceps, 202.
Motacilla capensis, 49, 70, 164, 168.
    Attacking a moth, 70.
    Pursuing species of *Acræa*, 70.
Moths flying to light, 122.
Mus coucha, 159.
—— *rattus*, 152, 159.
—— (*Isomys*) *pumilio*, 159.
Musca domestica, 244.
Museum at Cape Town, 5.
—— at Durban, 121.
—— at Port Elizabeth, 8.
Mutilla albistyla, sp. n., 211, 225.
—— *tetensis*, 211.
Mycalesis perspicua, 232.
Mygnimia belzebuth, sp. n., 211, 218.
—— *depressa*, sp. n., 211, 219.
—— *distanti*, sp. n., 211, 220.
—— *fallax*, sp. n., 211, 221.
Mylabris capensis, 199.
—— *gröndali*, 199.
—— *lunata*, 199.
—— *mixta*, 199.
—— *ophthalmica*, 92, 199.

Mylabris transversalis, 126, 199.
—— *tristigma*, 199.
Mylothris agathina, 234.
Myoxus murinus, 159.
Myriopoda, 181.
Myrmecocichla formicivora, 166.
Myrmeleon trivirgatus, 256.

Nanotragus scoparius, 159.
Nassunia bupaliata, 240.
Native commissioners and treatment of Kafirs, 98.
Nauphœta circumvagens, 258.
Nectarinia famosa, 166.
Nemognatha, 199.
Nephila transvaalica, sp. n., 91, 179, 180.
    Small colonies living in gigantic webs, 91, 179.
Neptis agatha, 233.
Nerium oleander, 18.
Neuroptera, 255.
    Habits &c., 255.
Neurosymploca agria, sp. n., 236.
—— *concinna*, 236.
Newcastle, 9, 119.
Nezara cupicola, 246.
—— *viridula*, 246.
Night in a wagon, 96.
Nilaus brubru, 167.
Nomophila noctuella, 153, 241.
Nucras tessellata, 174.
Numida coronata, 13, 72.
Nylstroom, 80.
Nysius noritius, sp. n., 247, 254.

Oceanic birds, 4.
Ochrophlebia ligneola, 259.
Ocypete megacephala, 180.
Œcanthos capensis, 258.
Œdalus acutangulus, 260.
—— *citrinus*, 260.
—— *marmoratus*, var., 260.
—— *nigro-fasciatus*, 260.
—— *plena*, 260.
Œdicnemus capensis, 169.

Oleanders, 18.
Oncopeltus famelicus, 247.
Oniticellus militaris, 191.
Onitis caffer, 191.
Onthophagus gazella, 191.
Ootheca modesta, sp. n., 206.
Ophisma croceipennis, 239.
—— pretoriæ, sp. n., 239.
Orthetrum fasciculata, 255, 256.
—— subfasciolata, 255, 256.
Orthoptera, 153, 257.
 Habits &c., 257.
 The food of many birds, 112.
Oryctes boas, 193.
Osteodes turbulentata, 240.
Otis afroides, 74, 168.
—— cærulescens, 74, 75, 168.
—— kori, 74, 128, 168.
 Food, 74.
 Maximum weight, 74.
Oxythyrea æneicollis, 194.
—— amabilis, 194.
—— cinctella, 194.
—— hæmorrhoidalis, 195.
—— marginalis, 194.
—— perroudi, 194.
—— rubra, 194.
Ozarba densa, 238.
—— dubitans, 238.

Pachnoda flaviventris, 92, 115, 126, 194.
—— leucomelana, 194.
Pachycnema tibialis, 192.
Pachymerus apicalis, 247.
Pachytylus lucasii, 260.
—— migratoroides, 71, 260.
 Immense swarm, 71.
Pais decora, 236.
Paliga infuscalis, 241.
Palpares caffer, 91, 256.
Palpopleura jucunda, 256.
—— lucia, 123, 256.
—— portia, 256.
Pamera proxima, 247.
Pamphila hottentota, 235.

Pangonia subfascia, 244.
Pantoleistes princeps, 247.
Papilio demoleus, 45, 123, 235.
—— morania, 123, 235.
Paracinema basalis, 260.
—— tricolor, 260.
Paramecocoris atomarius, 245.
—— ventralis, 240.
Parathespis galeata, 258.
Parga gracilis, 259.
Paroeme gahani, sp. n., 201, 202.
Partridges, 75, 112.
Parus afer, 167.
Pascoe, F. P., 200.
Passengers to South Africa; their sociological characteristics, 2.
Passer arcuatus, 168.
Passiflora, 18.
—— quadrangularis, 122.
Peach-brandy, 17.
Peach-trees, 17, 67.
Pedetes capensis, 75, 159.
Pelea capreolus, 159.
Pelopæus spirifex, 210.
Pentaplatarthrus natalensis, 40, 189.
Pentila tropicalis, 234.
Pernis apivorus, 165.
Petasia spumans, 259.
—— ——, var. ater (var. n.), 259.
Petovia dichroaria, 237.
—— marginata, 237.
Pezotettix capensis, 260.
Phalops flavocinctus, 191.
Pheropsophus fastigiatus, 188.
—— litigiosus, 40, 188.
 Effects of anal explosion, 40.
Philagathes lætus, 201.
Philonthus punctipennis, 189.
—— varians, 152, 189.
Phryneta spinator, 202.
Phrynobatrachus natalensis, 176.
Phyllocnema latipes, 201.
Phyllocrania paradoxa, 258.
Phyllodromia germanica, 153, 258.
Phymateus leprosus, 259.
—— morbillosus, 259.

Phymateus squarrosus, 259.
Physaliæ, 4.
Physorhynchus crux, 247.
―― *patricius*, 247.
Pieris gidica, 234.
―― *mesentina*, 234.
―― *severina*, 234.
Pietersburg, 95.
    Great advance in price of building-land, 95.
    In course of development, 95.
    Most German town in the Transvaal, 138.
    The Market town for Zoutpansberg, 95.
Piezia angusticollis, 188.
Piezodorus purus, 246.
Pine-apples, 121, 125.
Pionea africalis, 241.
Pirates conspurcatus, sp. n., 248, 255.
―― *lugubris*, 248.
Plæsiorrhina plana, 115, 126, 193.
Planema esebria, 123, 233.
Platypleura dirisa, 67, 248.
    Captured and eaten by Spiders, 67.
    Possibly pair during the breeding-season, 67.
Platypleura punctigera, 123, 248.
Plectropterus gambensis, 74, 169.
Pleonomus wahlbergi, 195.
Plinachtus falcatus, sp. n., 246, 251.
―― *pungens*, 246.
Plocepasser mahali, 167.
Plusia acuta, 238.
Pocock, R. I., 179, 181.
Podiceps capensis, 169.
Podontia nigrotessellata, 206.
Poliospiza tristriata, 168.
Polyclaeis cinerеis, 54, 200.
―― *equestris*, 54, 200.
    Impaled on spine of Acacia, 55.
    Victims of Spiders, 55.
Polydesma landula, 238.
Polygamy, 99.
Polyhirma macilenta, 50, 187.

Polysticta clarkii, 206.
Poophilus actuosus, 248.
Popillia bipunctata, 123, 192.
Porpoises, 4.
Port Elizabeth, 7.
    Botanic Gardens, 8.
    Climate, 7.
    Hospitable community, 7.
    Museum, 8.
Porthetis cinerascens, 259.
―― *griseus*, 259.
"Portuguese Man-of-War," 4.
Potgieter, Hermanus, Field-cornet, 81.
Potgieter's Rust, 81.
    Terrible visitation of fever in 1870, 81.
Pratincola torquata, 166.
Precis archesia, 233.
―― *ceryne*, 233.
―― *cloantha*, 46, 233.
―― *natalica*, 233.
―― *sesamus*, 233.
Pretoria (Town of), 17, 19, 47.
    Absence of auriferous deposits 59.
    Drainage and sanitary arrangements, 143.
    Geological features, 58.
    Most cosmopolitan town in the Transvaal, 138.
    Probable argentiferous wealth, 59.
    Social distinctions, 146.
    Vital statistics, 144.
Pretorius, Commandant, 82.
Prionemis hirsutus, sp. n., 211, 216.
Problem in the future of the Transvaal, 148.
Promeces viridis, 201.
Proscephaladeres obesus, 200.
―― *punctifrons*, 200.
Prospector, the, 85.
Protective resemblance, compound, 42.
Protoparce convolvuli, 18, 68, 153, 236.

T

Psammodes pierreti, 199.
—— *scabriusculus*, 199.
—— *striatus*, 54, 199.
    A prey to species of *Anthia* and *Manticora*, 54.
Psammophis sibilans, 176.
Psephus puncticollis, 195.
Pseudocolaspis sericata, 205.
Pseudonympha narycia, 232.
—— *vigilans*, 232.
Psiloptera calamitosa, 195.
—— *gregaria*, 195.
—— *viridimarginata*, 195.
Pternistes swainsoni, 168.
Pterochilus insignis, 210.
Pterocles gutturalis, 168.
Ptychophorus leucostictus, 194.
Puff-Adder, 173.
Pycnodictya adustum, 54, 260.
—— *obscura*, 260.
Pycnonotus layardi, 166.
Pyralis farinalis, 153, 241.
Pyrameis cardui, 39, 68, 153, 233.
Pyrgomantis singularis, 258.
Pyrgus diomus, 235.
—— *vindex*, 235.
Pyromelana oryx, 167.

Quaggas, 75.

Railway-construction, peculiarities of, 125.
Rains in 1890-91, 40, 46, 47, 48, 50, 53, 60, 86, 128.
Rana adspersa, 48, 173, 176.
—— *angolensis*, 176.
—— *natalensis*, 176.
Reduvius capitalis, sp. n., 247, 255.
—— *erythrocnemis*, 247.
—— *pulvisculatus*, sp. n., 247, 254.
—— *rapax*, 247.
—— *sertus*, sp. n., 247, 254.
Religious beliefs and customs of Magwambas, difficult to be explained or understood, 104.
Reptiles, 87.

Reptilia and Batrachia, 173.
Rhabdotis aulica, 194.
—— *semipunctata*, 194.
—— *sobrina*, 194.
Rhaphidopsis zonaria, 122, 202.
Rhinoceros, White, 5, 6.
Rhinoceros simus, 5, 6.
Rhopalocera, 231.
    Habits and distribution, 231, 232.
Rhynchota, 244.
    Habits &c., 244.
Rhyticoris terminalis, 246.
Richmond Road, Natal, 126.
Rocky plain resembling old ocean-bed, 97.
Roses, 18, 92, 126.
    Devastated in Natal by *Mylabris transversalis*, 126.
    Little visited by floral beetles, 92.
Routes from South-African Ports to the Transvaal, 6, 7.
Ruminants almost exterminated by Boers, 12.
Rutelidæ, 47, 192.
    Immense numbers under stones, 47.

Salamis anacardii, 123, 233.
Salix gariepina, 17.
Saprinus gabonensis, 190.
—— *natalensis*, 190.
Sarcophaga, sp., 244.
Saussure, Mons. Henri de, 212, 257, 260.
Saxicola familiaris, 166.
—— *monticola*, 166.
—— *pileata*, 166.
Scantius forsteri, 247.
Scaptelytra sulphureovittata, 195.
Scarabæidæ, 46, 190.
Scarabæus bohemani, 190.
—— *convexus*, 190.
—— *hottentottus*, 190.
—— *interstitialis*, 191.

Scarabæus nigroæneus, 190.
—— *savignyi*, 190.
Scarites rugosus, 188.
Schizonycha, sp., 192.
Schizorhis concolor, 112, 164, 166.
Scopus umbretta, 169.
Scotch colonist in Natal, his experience, 147.
Sebasteos galenus, 191.
Secretary-bird, 68, 164, 165.
    Capable of running miles without injury to legs, 69.
    Crop full of remains of orthopterous insects, 69.
Sennatocera, gen. n., 242.
—— *fuliginipuncta*, sp. n., 243.
Serica, spp., 192.
Serpentarius secretarius, 68, 164, 165.
Serricornia, 195.
Shark, 4, 8.
    Hammer-headed, 8.
Sharp, Dr. D., 189.
Sharpe, Dr. R. Bowdler, 164.
Shelley, Capt. G. E., 165.
Shrikes, their food, 55.
Siccia caffra, 237.
Silpha capensis, 189.
—— *pernix*, 189.
Silvius denticornis, 244.
Sinoxylon conigerum, 198.
Sitagra caffra, 167.
Smuggling in the Transvaal, 11.
Snakes not abundant, 87.
Snow, 48.
Social distinctions in Pretoria, 146.
Solpuga chelicornis, 179.
    Attacked by birds, 179.
Solpugæ, 179.
Species found both in England and the Transvaal, 67, 152.
Spelonken, 96.
Sphæroderma indica, 206.
Sphærotherium obtusum, 181.
Sphenophorus, sp., 200.
Sphex nigripes, 210.

Sphingomorpha monteironis, 239.
Spiders, 55, 67, 91, 179.
Spilocephalus distanti, sp. n., 208.
—— *viridipennis*, 206, 207.
Spilophorus plagosus, 194.
Spined and hard-wooded trees, survival of, in the struggle for existence, 43, 44.
Spirostreptus meinerti, 181.
—— *transvaalicus*, sp. n., 179, 181.
—— (*Odontopyge*) *pretoriæ*, sp. n., 183.
Spreo bicolor, 164, 167.
Spring in the Transvaal, 45, 46.
Standerton, 118.
Staphylinus hottentottus, 189.
Starvation, no, in South Africa, 139.
Steganocerus multipunctatus, 245.
Sternocera orissa, var., 91, 195.
    Habits and singular mode of capture, 92.
Sterrha sacraria, 153, 240.
Stewart, Prof. C., 157.
Store- and Canteen-keepers, 130.
Strix flammea, 165.
Strongylium, sp., 199.
Sympetrum fonscolombii, 256.
Synagris mirabilis, 210.
Syntomis khulweinii, 236.
Syrichthus verus, 193.

Tabanus socius, 244.
Tanning as practised by the Boers, 77.
Taphronota porosa, 259.
Tarache caffraria, 238.
Tarucus telicanus, 233.
Taurotagus klugii, 201.
Telephonus senegalus, 167.
Tephræa morosa, 194.
Tephrocorys cinerea, 168.
Teracolus achine, 234.
—— *eris*, 91, 234.
—— *evenina*, 91, 234.
—— *gavisa*, 234.
—— *omphale*, 234.
—— *phlegetonia*, 234.

Teracolus subfasciatus, 112, 234.
—— *theogone*, 234.
—— *vesta*, 113, 234.
Teracotana submacula, 237.
Terias brigitta, 39, 112, 234.
—— *zoë*, 112, 234.
Termes, sp., 48.
—— *angustatus*, 49, 256.
Termites devoured by Batrachians, 48.
—— —— by Dog, 50.
—— —— by the Cape Wagtail, 49.
Tetradia fasciatocollis, 202.
Tetragonoderus bilunatus, 188.
Theology of the Boers, 23, 26, 33.
Thomas, Oldfield, 157.
Tibicen carinatus, 248.
—— *undulatus*, 248.
Timber, want of, in the Transvaal, 78
Tinea, sp., 243.
—— *vastella*, 243.
Tithoes confinis, 201.
Tmetonota sabra, 260.
Toads, 87.
Trachynotus angulatus, 199.
Trachyphonus cafer, 166.
Traders with the Boers, 130.
Tragiscoschema amabilis, 202.
Tragocephala sulphurata, sp. n., 202.
Tramea basilaris, 54, 256.
Transvaalia, gen. n., 247, 253.
—— *lugens*, sp. n., 247, 253.
Tree-ferns, 121.
Triænogenius corpulentus, 188.
Trianeura fulvescens, 236.
Trichostetha placida, 123, 194.
Trigonopus, sp., 199.
Tritophidia annulata, 260.
Trimen, Mr. R., 5, 42, 231.
Tritæa lacunalis, 241.
Trithemis dorsalis, 256.
—— *sanguinolenta*, 112, 256.
Trochalus, sp., 192.
Tropicorypha corticina, 245.
Turdus litsitsirupa, 166.
Turtur senegalensis, 168.

United South-African Confederacy, 133.
Upupa africana, 166.
Urolestes melanoleucus, 167.
Urota sinope, 122, 238.

Varanus niloticus, 75, 173, 174.
  Food, 87.
Veld, 8.
Veterna patula, sp. n., 246, 250.
—— *pugionata*, 246.
—— *sanguineirostris*, 246.
—— *subrufa*, 246.
Vidua ardens, 167.
—— *principalis*, 167.
Vipera arietans, 173, 176.
Volksrust, 11, 129.
Vultures, 13, 69.
  Maximum weight, 69.
  Mode of flight, 70.
  Their probable greater scarcity of food, 70.

Wallace, A. R., 41.
Walsingham, Right Hon. Lord, 235.
Warm baths, 79.
Warren, W., 235.
"Waterberg Flats," 79.
Waterberg, scenery in, 80.
Waterfal River, 129.
Waterhouse, C. O., 190.
Weather, severe, and losses in live-stock, 48.
White Ants, 48.
White-bellied Crow, 90.
Willow trees, 17, 19, 67.
Winter or dry season in the Transvaal, 9, 10, 18, 39, 73, 111, 114.
Women, Kafir, 99.
  Female infanticide unknown, 99.
  Polygamy a useful institution, 100.
  Position of the Kafir women compared with those of the lowest classes in England, 100.

Wood-bush, 91.
    Contrast of its zoology with that of the bare veld, 91.

Xanthospilopteryx superba, 91, 236.
Xiphocera, 260.
—— *distanti*, sp. n., 260, 261.
—— *picta*, sp. n., 260, 261.
Xylocopa inconstans, 210.
Xystrocera globosa, 201.

Ypthima asterope, 232.

Zeritis orthrus, 234.
Zinnias, 18.
Zonitis eborina, 198, 199.
Zonurus cordylus, 174.
Zophosis angusticostis, 190.
—— *punctulata*, 199.
Zosterops virens, 106.
Zoutpansberg, 113.
    Great agricultural future, 113.
    Warm air and rich vegetation, 113.
Zygæna, sp., 8.

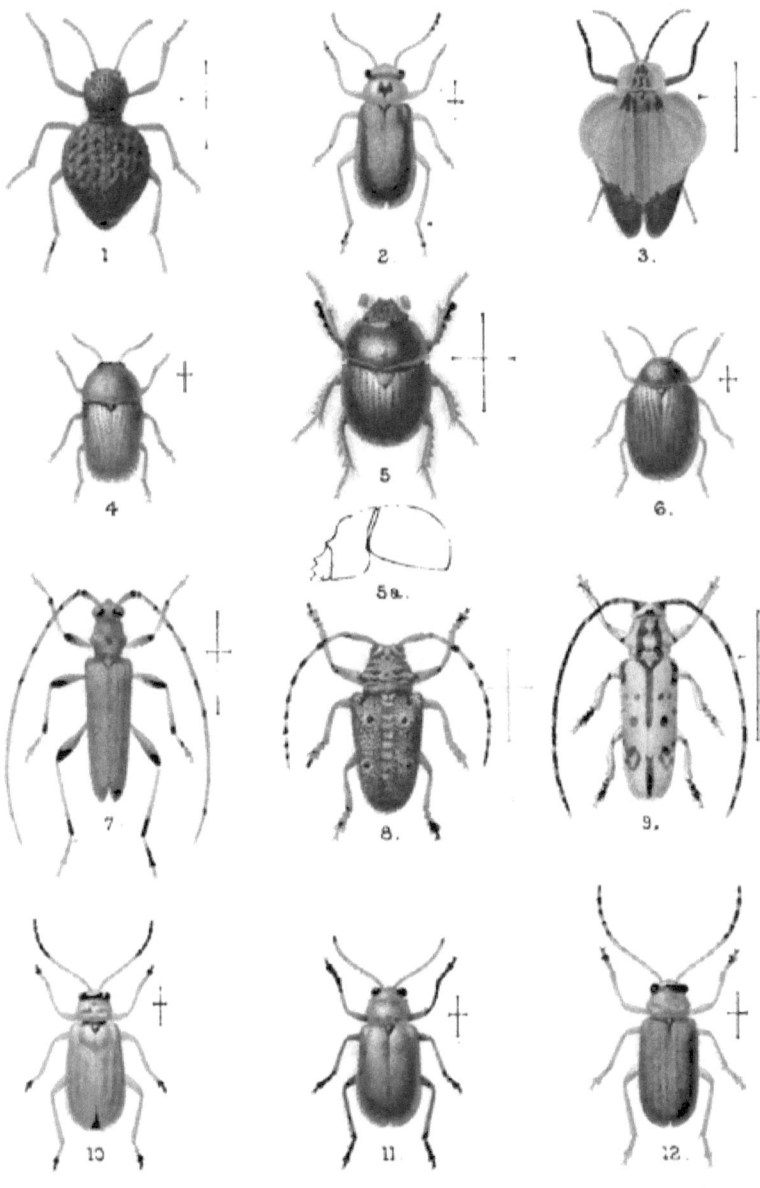

W. Purkiss del.   Mintern Bros. Chromo.

1 Amiantus undosus.   5 Bolboceras batesii.   9 Tragocephala sulphurata.
2 Hedybius amœnus.   6 Menius distanti.   10 Ænidea pretoriæ.
3 Lycus distanti.   7 Paroeme gahani.   11 Oothœca modesta.
4 Achœnops facialis.   8 Crossotus klugii.   12 Spilocephalus viridipennis.

R.H Porter, Publisher, London.

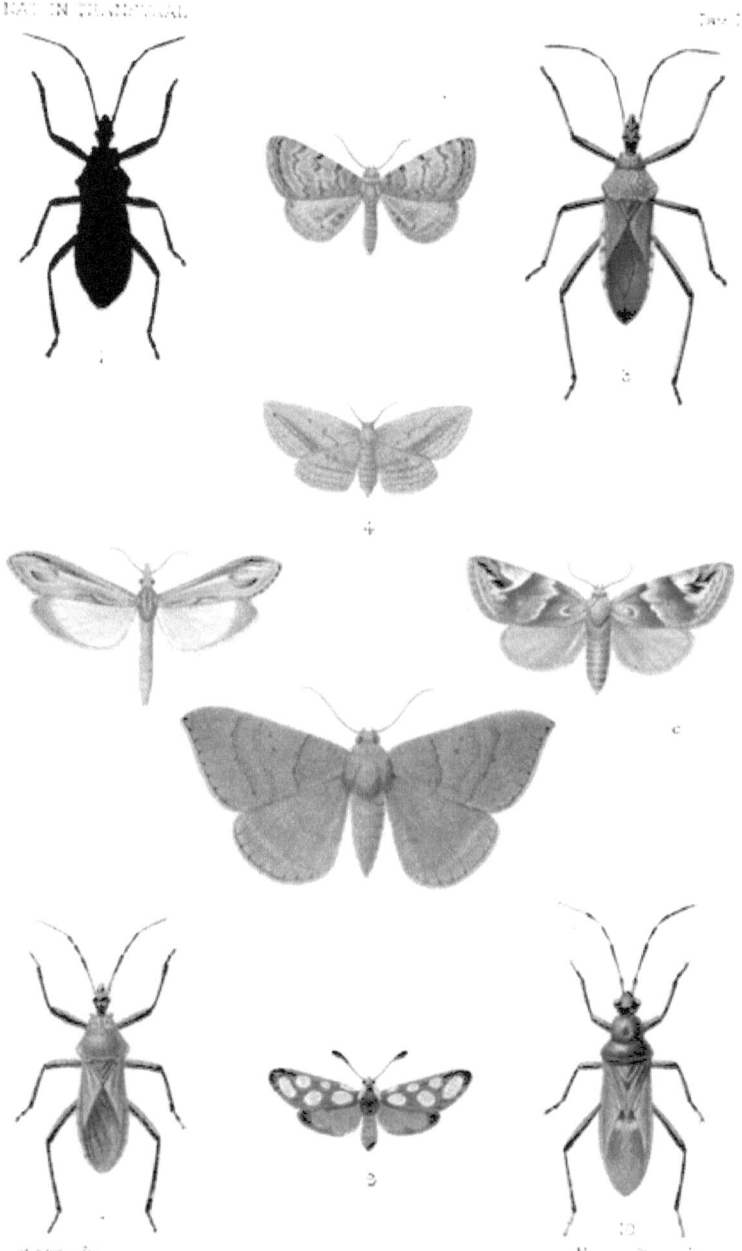

NAT. IN TRANSVAAL. Tab. III.

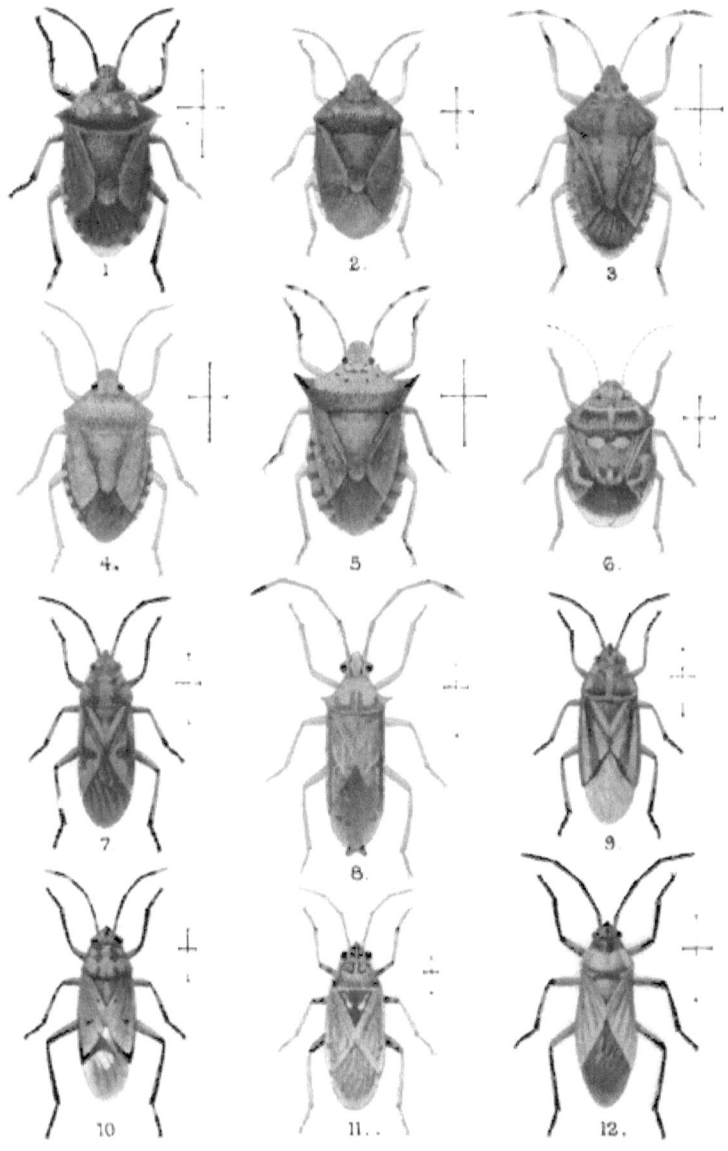

W Purkiss del. Mintern Bros. Chromo

1. Cimex figuratus, var.
2. Holcostethus obscuratus
3. Halyomorpha capitata
4. Halyomorpha pretoriæ
5. Veterna patula
6. Antestia transvaalia
7. Lygæus planitiæ
8. Plinachtus falcatus
9. Lygæus desertus
10. Lygæus campestris
11. Nysius novitius
12. Transvaalia lugens

R H Porter, Publisher, London

NAT. IN TRANSVAAL. Tav. IV

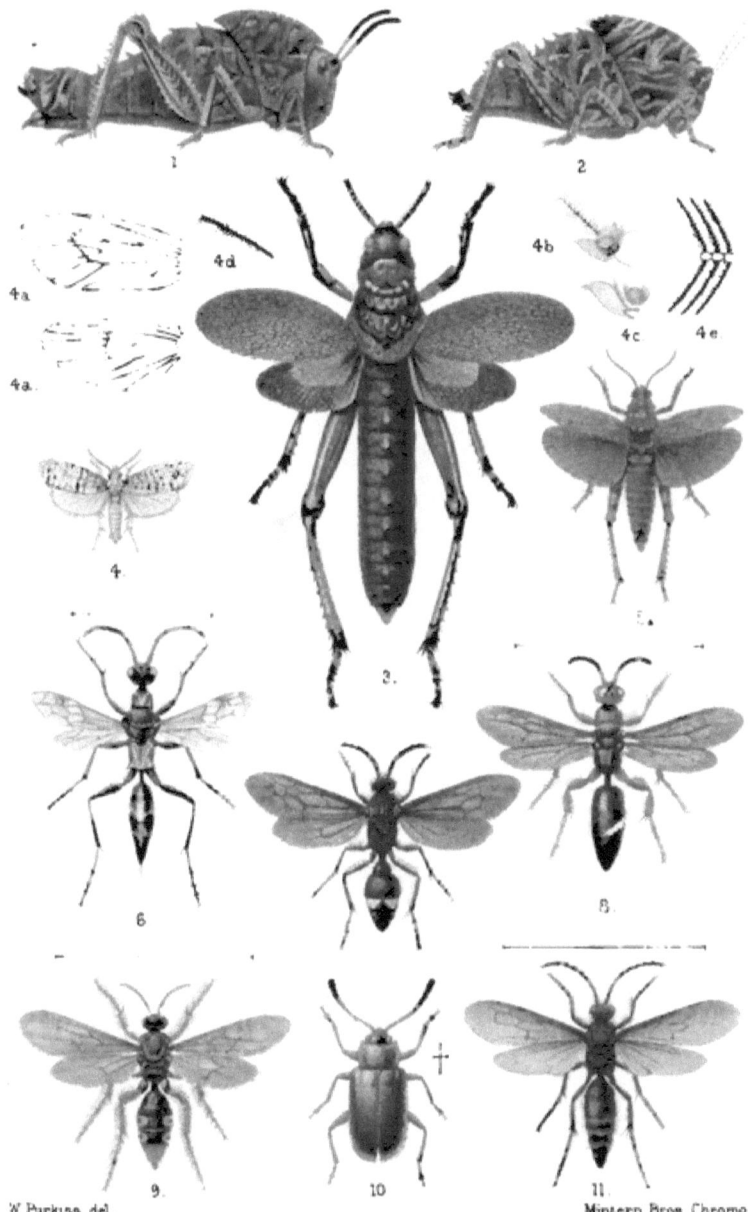

W Purkiss del.　　　　　　　　　　　　　　Mintern Bros. Chromo.
  1. Xiphocera distanti.　　4,a,b,c,d,e, Sematocera gen charact. 8. Mesa diapherogamia.
  2. Xiphocera picta.　　　5. Chrotogonus meridionalis　　9. Discolia præstabilis
  3. Petasia spumans var ater　6. Ampulex nigrocœrulea.　　10. Ancylopus fuscipennis.
  4. Sematocera fuliginipuncta. 7. Mutilla albistyla.　　　　11. Discolia præcana.
　　　　　　　　R. H. Porter, Publisher, London.

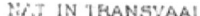

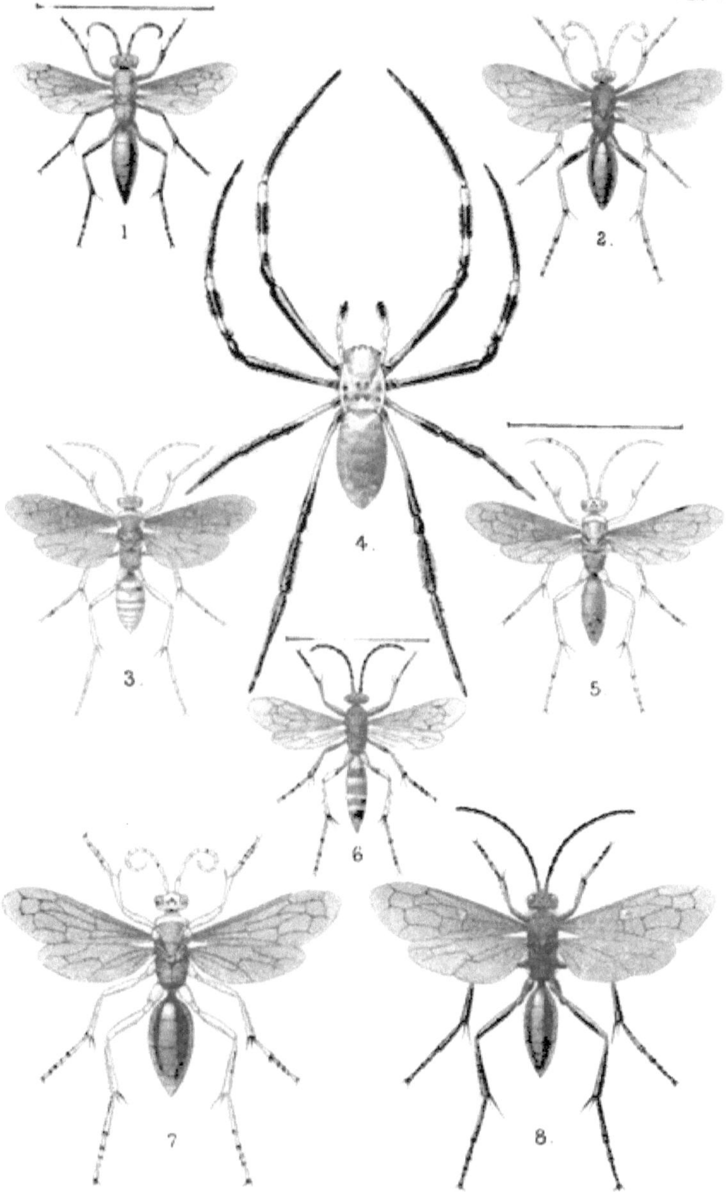

Porrosa del. et lith. Mintern Bros. imp.

1. Homonotus cœrulans.
2. Cyphononyx antennata.
3. Priocnemis hirsutus.
4. Nephila transvaalica.
5. Mygnimia fallax.
6. Homonotus pedestris.
7. Mygnimia distanti.
8. Mygnimia belzebuth.

R. H. Porter, Publisher, London.

www.ingramcontent.com/pod-product-compliance
Lightning Source LLC
Chambersburg PA
CBHW022020240426
43667CB00042B/1003